Edward W. Janson

British Beetles

Transferred from Curtis's British Entomology

Edward W. Janson

British Beetles
Transferred from Curtis's British Entomology

ISBN/EAN: 9783337144043

Printed in Europe, USA, Canada, Australia, Japan

Cover: Foto ©berggeist007 / pixelio.de

More available books at **www.hansebooks.com**

BRITISH BEETLES.

TRANSFERRED FROM

CURTIS'S BRITISH ENTOMOLOGY.

WITH DESCRIPTIONS

BY E. W. JANSON, ESQ.

FORMERLY SECRETARY OF THE ENTOMOLOGICAL
SOCIETY.

LONDON:
BELL AND DALDY, 186 FLEET STREET.
1863.

BRITISH BEETLES.

PLATE I.

Fig. 1. CICINDELA HYBRIDA, *Linn.* [*F.* Cicindelidæ. *G.* Cicindela, *Linn.* TIGER BEETLES.] (C. sylvicola, *Curtis,* but not of *Dejean.*)

Above, brassy brown, with a greenish hue : sides of head, margins of thorax, suture and external margin of elytra, second and third joints of antennæ, and base of thighs and of tibiæ, bright coppery red ; basal joint of antennæ, tarsi and apex of thighs and tibiæ, purplish green : labrum, base of mandibles, a humeral and apical lunule, and a central broad, more or less deflexed, band on each elytron, white. Beneath, with a long white pubescence, green : the breast and sides of thorax, coppery red. Length, 6–7 lines.

Abundant on the coast of Lancashire and North Wales, but examples presenting the bright green colour of the one figured appear to be of exceedingly rare occurrence.

C. sylvicola, Dej., is an Alpine species not hitherto detected in Britain.

Fig. 2. LEISTUS FULVIBARBIS, *Dej., Curtis.* [*F.* Carabidæ. *G.* Leistus, *Fröhlich.*]

Brownish black, with a faint bluish tint ; mouth, antennæ, and legs red ; abdomen beneath reddish brown. Thorax heart-shaped, convex ; hinder angles rectangular, acute ; anterior margin and base with numerous large punctures. Elytra striate, the striæ punctured. Length, 3–3½ lines.

Common in damp woods and swamps, beneath fallen leaves, and at the roots of grass.

Fig. 3. NEBRIA LIVIDA, *Linn., Curtis.* [*F.* Carabidæ. *G.* Nebria, *Latreille.*]

Shining brownish black ; mouth, antennæ, legs, thorax—except the front and hind margins— and outer margins of elytra, pale yellow ; beneath, reddish brown ; apex of abdomen, red. Thorax transverse, narrowed behind. Elytra striate, the striæ punctured, the third interstice with three or four large punctures contiguous to the third stria. Length, 7–8 lines.

Abundant at Bridlington Quay, near Scarborough, in the interstices of the argillaceous cliffs.

Fig. 4. CYCHRUS ROSTRATUS, *Linn., Curtis.* [*F.* Carabidæ. *G.* Cychrus, *Fabricius.*]

Deep black, above slightly shining, beneath glossy. Thorax with the sides reflexed, especially posteriorly, and with a deep transverse impression behind, its surface coarsely chagrined. Elytra ovate, chagrined like the thorax, each elytron usually presenting three faint, raised, longitudinal, interrupted lines. Length, 8–9 lines.

Common in sand-pits, in moss at roots of trees in woods, and a frequent visitor at night to sugar spread on the trunks of trees to allure Lepidoptera.

Fig. 5. CALOSOMA SYCOPHANTA, *Linn., Curtis.* [*F.* Carabidæ. *G.* Calosoma, *Weber.*]

Above, deep blue-black, elytra golden green, with a fiery, coppery red tint at the sides ; beneath and legs, black. Head and thorax thickly covered with confluent punctures. Elytra with sixteen

B   1

PLATE I.—*Continued.*

punctured striæ, the interstices transversely wrinkled, the fourth, eighth, and twelfth with a row of five or six large punctures. Length, 10–15 lines.

Occasionally found—sometimes floating on the sea at a considerable distance from the shore—on the coasts of Norfolk, Suffolk, Kent, the Isle of Wight, Devonshire, and Ireland ; and, probably, not truly indigenous.

Fig. 6. CARABUS VIOLACEUS, *Linn. var. c.* [*F.* Carabidæ. *G.* Carabus, *Linn.*] (*C.* exasperatus, *Dufts, Dej., Curtis.*)

Oblong, black ; sides of thorax and of elytra blue, violet, or coppery. Head and thorax finely confluently punctured, the latter quadrate, with its lateral margins reflexed, and its posterior angles produced and deflexed. Elytra elongate ovate, thickly granulate, the granules united, and forming distinct, but more or less interrupted and irregular elevated longitudinal lines. Length, 12–13 lines.

Captured by Mr. Curtis in the Isle of Portland.

The typical form, in which the granules on the elytra are finer, and present scarcely any trace of lines, is abundantly distributed, being found in sand-pits, fields, and gardens, and at the roots of trees in woods, throughout the country.

Fig. 7. PELOPHILA BOREALIS, *Payk., Curtis.* [*F.* Caribidæ. *G.* Pelophila, *Dejean.*]

Above, brassy black, the sides of the thorax, and the elytra with a greenish or coppery tint ; beneath, black. Head longitudinally wrinkled on each side near the eyes, and with two oblong, shallow foveæ anteriorly between the antennæ, obsoletely transversely wrinkled behind. Thorax wider than the head, transverse, narrowed posteriorly, thickly and coarsely punctured in front and behind, lateral margins reflexed. Elytra nearly twice as wide as the thorax, oblong, very shallowly striate, the striæ minutely punctured, the third and fifth interstices with a series of large shallow impressions variable in number. The legs are generally rusty red, sometimes brown, occasionally nearly or wholly black. Length, 5–6½ lines.

Sandy shores of Lough-Neagh, Ireland ; and in the Orkney and Shetland Isles.

Fig. 8. BRACHINUS SCLOPETA, *Fab., Curtis.* [*F.* Brachinidæ. *G.* Brachinus, *Weber.* BOMBARDIER BEETLES.]

Oblong, convex, rusty red ; elytra, bright blue ; the suture, from the base to within about one-third of the apex, rusty red. Length, 2¼–3¼ lines.

Very rare in Britain. The following localities have been recorded :—Devonshire ; Southend, Essex ; Hastings, Sussex ; and Norfolk.

Fig. 9. NEBRIA GYLLENHALII, *Schœnh.* [*F.* Carabidæ. *G.* Nebria, *Latreille.*] (Helobia Gyllenhalii, *Curtis.*)

Black. Head with two shallow foveæ in front, and a small shallow impression on the crown. Thorax heart-shaped, transverse, constricted behind ; the posterior angles acutely rectangular, the lateral margins reflexed. Elytra oblong, their sides nearly parallel, deeply striate, the striæ more or less distinctly punctured ; a row of from three to five large punctures on the third interstice contiguous to the third stria. Legs black : tarsi reddish, or pitchy red, with the tips of the thighs darker, or entirely rusty red. Length, 4–5 lines.

An abundant species in the mountains of North Wales, the north of England, Ireland, and Scotland.

PLATE II.

Fig. 10. DRYPTA DENTATA, *Rossi.* [*F.* Dryptidæ. *G.* Drypta, *Fabricius.*] (D. emarginata, *Fab., Curtis.*)

Bright green, or blue with a greenish tint ; clothed with a short, sub-erect, amber yellow pubescence ; labrum, mandibles, palpi, legs, and antennæ red, the latter with the apex of the first and third joints, and a patch on the second, dusky. Head and thorax thickly and coarsely punctate. Elytra deeply striate, the striæ deeply punctate, interstices flat, punctured. Length, 4 lines.

Exceedingly local, and rare. Lyme Regis, Dorset ; Faversham, Kent ; Hastings, Sussex ; Luccombe, Isle of Wight ; and Alverstoke, Hants.

Fig. 11. LEBIA TURCICA, *Fab., Curtis.* [*F.* Lebiidæ. *G.* Lebia, *Latr.*]

Black ; mouth, antennæ, thorax, scutellum and legs red ; elytra with a large triangular pale yellow humeral patch extending to the second stria, the striæ fine but deep, minutely punctate, interstices obsoletely punctate, the third with two large punctures contiguous to the third stria : abdomen with a pitchy red patch in the middle beneath. Length, 2–2½ lines.

Taken by the late Dr. Leach in Oakhampton Park, Devon.

Fig. 12. LEBIA CYANOCEPHALA, *Linn.* [*F.* Lebiidæ. *G.* Lebia, *Latr.*] (Lamprias cyanocephalus, *Curtis.*)

Shining blue, sometimes with a purple tint ; basal joint of antennæ, thorax, and legs red ; rest of antennæ, apex of thighs, and tarsi black. Elytra finely punctate—striate, interstices conspicuously punctate. Length, 2¼–3½ lines.

Appears partial to the chalk, and occurs in many places in the south of England, beneath stones, in moss, and by brushing. Darenth, Kent ; Reigate and Micklebam, Surrey, &c.

Fig. 13. POLYSTICHUS VITTATUS, *Brullé.* [*F.* Dryptidæ. *G.* Polystichus, *Bonelli.*] (P. fasciolatus, *Fab., Curtis,* but not of *Rossi.*)

Above, pitchy brown, with a long yellow pubescence ; a central longitudinal stripe on each elytron, antennæ and legs ferruginous. Head and thorax thickly and deeply punctate. Elytra deeply striate, the striæ finely punctate, interstices very thickly and finely punctate, with a row of large punctures between the outer stria and the margin. Length, 4–4½ lines.

Excessively local, but occasionally abundant. Southwold, Suffolk ; Cley, Norfolk ; Sandown, Isle of Wight ; and near Hastings, Sussex.

Fig. 14. ODACANTHA MELANURA, *Linn., Curtis.* [*F.* Odacanthidæ. *G.* Odacantha, *Paykull.*]

Head and thorax bright bluish green, the latter sub-cylindrical, much narrower than the head, transversely wrinkled, sparsely and coarsely punctate, with a fine central longitudinal impressed line. Mandibles red ; palpi and antennæ black, the latter with the three basal joints and the extreme base of the fourth yellowish red. Elytra yellowish red, with a bluish-black apical patch, finely striate-punctate, interstices smooth. Abdomen bluish black ; breast, yellowish red ; legs, yellowish red ; tips of thighs, black ; tarsi, pitchy. Length, 3¼ lines.

3

PLATE II.—*Continued.*

Abundant, in the spring and autumn, in the fenny districts of Huntingdonshire, Norfolk, and Cambridge; also near Swansea; in Surrey, near Three Bridges and Earlswood Common; in the stems of the Reed-mace (*Typha latifolia*); it has likewise occurred in the vicinity of the metropolis, near Hammersmith.

Fig. 15. CYMINDIS VAPORARIORUM, *Linn.* [*F.* Lebiidæ. *G.* Cymindis, *Latr.*] (Tarus basalis, *Curtis.*)

Reddish brown or brownish black, pubescent; antennæ and parts of the mouth pitchy red. Head and thorax rather thickly and coarsely punctate; the latter heart-shaped, with a faint central longitudinal channel, its posterior angles acute and prominent. Elytra elongate ovate, rather convex, punctate-striate, interstices thickly and deeply punctate, the base red brown. Legs, pitchy red. Length, 4–4½ lines.

Not uncommon in the mountainous districts of Wales and Scotland.

Fig. 16. DEMETRIAS MONOSTIGMA, *Leach,** Curtis.* [*F.* Lebiidæ. *G.* Demetrias, *Bonelli.*]

Reddish yellow. Head black, shining; mandibles, pitchy red. Thorax heart-shaped, with a very conspicuous central longitudinal channel, its posterior angles slightly prominent. Elytra faintly punctate-striate, with a common pitchy-black, ill-defined patch at the apex within. Length, 2–2¼ lines.

Formerly abundant in the fens of Huntingdonshire and Cambridgeshire; common at the roots of grass, &c., near Deal, and in other places on the south and west coasts in the spring and autumn.

Fig. 17. METABLETUS OBSCUROGUTTATUS, *Duftschmidt.* [*F.* Lebiidæ. *G.* Metabletus, *Schmidt-Goebel.*] (Dromius spilotus, *Dejean, Curtis.*)

Brownish black with a slight brassy hue. Head smooth; antennæ black; pitchy red at the base. Thorax short, sub-quadrate, narrowed behind, with a distinct central longitudinal channel, its posterior angles obtuse. Elytra very faintly striate, with an ovate spot, frequently indistinct, on the shoulders, occasionally another smaller one near the apex, and the reflexed margin brownish yellow. Legs pitchy red, the thighs darker. Length, 1¼–1½ lines.

Not common, although widely distributed. Near London, at Hornsey, at roots of grass on shady banks; Holme Fen, Hunts; Isle of Wight; Tunbridge Wells; near Reigate, Surrey, &c. &c.

Fig. 18. CLIVINA COLLARIS, *Herbst, Curtis.* [*F.* Scaritidæ. *G.* Clivina, *Latr.*]

Head reddish brown, smooth, with a central impressed line on the forehead, and two large punctures behind the eyes; antennæ, mandibles, and palpi, red. Thorax, pitchy brown, subquadrate, slightly narrowed anteriorly, with a transverse arcuate impressed line in front, and a conspicuous central longitudinal channel. Elytra red; the suture frequently pitchy black; deeply striate, the striæ closely and rather finely punctate; interstices flat, the third with two or three large punctures contiguous to the third stria. Legs red. Length, 2½ lines.

Local, but widely, although sparingly, distributed throughout England, and occasionally found in Scotland.

* In Samouelle's *Entomologist's Useful Compendium.* p. 156, 2 (1819). *D.* unipunctatus, *Germ.* Ins. Spec. Nov. I, I, 2 (1821).

4

PLATE III.

Fig. 19. FERONIA (LYPERUS) ATERRIMUS, *Payk.* [*F.* Feroniidæ. *G.* Feronia, *Latr.* *s-G.* Lyperus, *Chaudoir.*] (Omaseus aterrimus, *Curtis.*)

Deep shining black. Head much narrower than the thorax; smooth, with two deep curved impressions between the antennæ. Thorax sub-quadrate, scarcely narrowed behind, slightly rounded at the sides, its lateral margins reflexed, posterior angles rounded, disk convex, a deep finely rugulose fovea on each side at the base. Elytra a trifle wider than the thorax, elongate, finely punctate-striate; interstices flat, smooth, the third with three large impressions, the anterior of which is contiguous to the third stria, the two posterior to the second stria. Length, 6½-7 lines. Formerly abundant at Whittlesea Mere, and in the fens of Cambridgeshire and Norfolk.

Fig. 20. FERONIA (PTEROSTICHUS) ELONGATUS, *Curtis.* [*F.* Feroniidæ. *G.* Feronia, *Latr.* *s-G.* Pterostichus, *Bonelli.*] (Pterostichus pyrenæus? *de Chaudoir.*)

Narrow, elongate, shining black. Head large, nearly as wide as the thorax; antennæ, brownish, the three basal joints black. Thorax broader than long, narrowed behind, posterior angles acute, with a conspicuous central longitudinal channel, and an elongate fovea on each side at the base. Elytra scarcely wider than the thorax, elongate, their sides parallel, sinuate-truncate at the apex; the sutural angle acute; finely striate, the striæ minutely, remotely, and obscurely punctate; interstices flat, smooth, the third with five large impressed points, of which the anterior is contiguous to the third, the four posterior to the second stria. Length, 7½ lines.

Nearly allied to Feronia (Pterostichus) Lasserei, Fairm., but readily distinguished by its narrower form, and its elongate, depressed, parallel, sub-truncate elytra.

Introduced into the British list upon the authority of an example, extant in the British Museum, supposed to have been taken, many years ago, in Devonshire, by Dr. Leach.

Fig. 21. POGONUS LURIDIPENNIS, *Germar.* [*F.* Feroniidæ. *G.* Pogonus, *Dejean.*] (P. Burrellii, *Curtis.*)

Head and thorax bright green, with a brassy or coppery tint; palpi, antennæ, legs, and elytra, pale yellow; the latter occasionally with a dark, ill-defined patch on the disk. Head conspicuously narrower than the thorax; smooth behind; finely chagrined anteriorly and at the sides, with two broad longitudinal impressions between the eyes. Thorax much broader than long, rounded at the sides, narrowed behind; the base depressed and ruggedly punctate; posterior angles nearly rectangular; foveæ shallow, bounded externally by a little ridge; disk convex transversely wrinkled, dorsal channel distinct. Elytra elongate-ovate, finely punctate-striate; interstices flat, the third with three impressions, the two anterior contiguous to the third stria, the posterior to the second stria. Underside brassy green. Length, 3¼-3¾ lines.

First discovered in Britain near Salthouse on the coast of Norfolk. Taken, also, near Lymington; and abundantly near Sheerness, on the muddy margins of pools of salt water.

Fig. 22. DYSCHIRIUS NITIDUS, *Dejean.* [*F.* Scaritidæ. *G.* Dyschirius, *Fabricius.*] (D. inermis, *Curtis.*)

Brassy black, with a slight bluish or greenish tint. Head usually smooth, bi-dentate in front, forehead with a deep transverse impressed line, which is sometimes interrupted in the middle, a deep furrow on each side, the space intervening between it and the eye forming an elevated ridge, a broad, shallow depression in the centre, and one or two indistinct transverse striæ behind; mandibles, palpi, and base of antennæ, pitchy red. Thorax ovate, convex, longer than broad, its greatest width a little behind the middle, with an arcuate impressed line in front, the space enclosed between it and the anterior margin with fine closely set longitudinal striæ, central longitudinal channel deep. Elytra oblong ovate, sub-truncate at the base, shoulders moderately rounded, with eight dorsal striæ, the first and second uniting at the base in a deep round impression, the striæ of equal depth from base to apex, but becoming gradually shallower towards the sides, distinctly punctured at the base, interstices smooth, the third with two or three large punctures contiguous to the third stria. Legs reddish-brown; of the anterior the thighs robust, black, the tibiæ wide, their external margin unarmed. Length, 2-2¼ lines.

Preston Marsh, and Lytham, Lancashire; and, it is said, near Yarmouth.

5

PLATE III.—*Continued.*

Fig. 23. **MISCODERA ARCTICA,** *Payk.* [*F.* Broscidæ. *G.* Miscodera, *Eschscholtz.*]
(Leiochiton Readii, *Curtis.*)

Shining bluish or greenish-black; parts of the mouth, antennæ, and legs, red. Head smooth, with two longitudinal impressions between the eyes, and a transverse impression in front. Thorax globose, smooth, as long as wide, constricted and punctate behind, the sides rounded, with a faint central longitudinal impressed line. Elytra ovate, punctate-striate, the striæ obliterated at the apex and at the sides. Beneath brown, apex of abdomen pitchy red. Length, 3 lines.
Inhabits the mountains of North Wales and Scotland, and the elevated moors of Yorkshire.

Fig. 24. **FERONIA (STEROPUS) ÆTHIOPS,** *Panzer.* [*F.* Feroniidæ. *G.* Feronia, *Lat.* *s-G.* Steropus, *Megerle.*] (Steropus concinnus, *Curtis.*)

Black, shining. Head smooth with a broad, shallow impression on each side in front. Thorax transverse, its greatest width a little before the middle, rounded at the sides, narrowed behind, disk convex, with a distinct central longitudinal furrow, posterior angles rounded, on each side at the base a large ruggedly punctate fovea. Elytra ob-ovate, convex, deeply striate, the striæ impunctate, interstices convex, the third with two or three large punctures contiguous to the third stria. The male has the penultimate segment of the abdomen armed beneath in the centre with an acute tubercle, and the apex of the terminal segment with a shallow fovea. Length, 5-5½ lines.
Not uncommon on the mountains of North Wales, Cumberland, and Scotland.

Fig. 25. **PATROBUS SEPTENTRIONIS,** *Dejean.* [*F.* Feroniidæ. *G.* Patrobus, *Dejean.*] (P. alpinus, *Curtis.*)

Pitchy black. Head smooth, with a deep longitudinal punctate furrow on each side between the eyes, and a deep punctate transverse impression behind; palpi pitchy red, antennæ, dark brown. Thorax convex, transverse, narrowed behind, the sides rounded in front, posterior angles acutely rectangular, a triangular impressed sparsely punctate space in front, a large coarsely punctate fovea on each side behind. Elytra oblong ovate, the shoulders rather prominent, striate, the striæ thickly punctate, interstices flat, the third with three large punctures. Wings ample. Beneath with the prothorax somewhat thickly and deeply punctate, flanks of the metathorax obsoletely punctate. Thighs pitchy black, tibiæ and tarsi pitchy red. Length, 4¼-4¾ lines.
An alpine species, found on Ben Lomond and other mountains of Scotland.

Fig. 26. **DIACHROMUS GERMANUS,** *Linn.* [*F.* Harpalidæ. *G.* Diachromus, *Erich.*]
(Ophonus germanus, *Curtis.*)

Above clothed with a dense short pubescence, thickly and deeply punctate. Head reddish yellow; antennæ, with the exception of the four basal joints, pitchy brown. Thorax blue or bluish green, its lateral margins narrowly red, heart-shaped, posterior angles acute, with a deep elongate impression at the base on each side. Scutellum black. Elytra ovate, reddish yellow, with a common bluish green patch at the apex within, striate, the striæ usually impunctate. Beneath, except the head, deep shining black. Legs testaceous. Length, 4-4½ lines.
Very rare in Britain. It has occurred at Kingsbridge, Hastings, and Deal.

Fig. 27. **HARPALUS LATUS,** *Linn.,* var. [*F.* Harpalidæ. *G.* Harpalus, *Latr.*]
(Harpalus ruficeps, *Curtis.*)

Pitchy black, sides of thorax, palpi, antennæ, and legs, and occasionally the head and lateral margins of the elytra brownish red. Head wide. Thorax transverse, quadrate, scarcely perceptibly narrowed behind, with a distinct central longitudinal furrow, posterior angles rectangular with their extreme summits rounded, base thickly rugulose punctate, with a shallow elongate fovea on each side. Elytra ovate, slightly emarginate near the apex, deeply striate, the striæ impunctate, interstices smooth. Length, 3¾-4¼ lines. Common throughout the country.

PLATE IV.

Fig. 28. LICINUS DEPRESSUS, *Payk., Curtis.* [*F.* Licinidæ. *G.* Licinus, *Latr.*]

Black, sub-opaque. Head sparsely and finely punctate. Thorax convex, a little wider than long, with a conspicuous central longitudinal furrow, slightly rounded at the sides, narrowed behind, emarginate at the base, the lateral margins reflexed posteriorly, posterior angles rounded, very thickly and somewhat coarsely punctate. Elytra depressed, a little wider than the thorax, finely striate, the striæ finely punctate, interstices flat, thickly punctate. Length, 4–4½ lines.

In Britain this species appears to be confined to the chalk, and to be distributed over a very limited area. It has occurred, occasionally in some plenty, at Box Hill, Mickleham, and near Reigate, Surrey; at Dover, and near Canterbury, Kent; and near Brighton, Sussex.

Fig. 29. FERONIA (POECILUS) LEPIDUS, *Fab.* [*F.* Feroniidæ. *G.* Feronia, *Latr.* *s-G.* Poecilus, *Bonelli.*] (*P.* lepidus, *Curtis.*)

Black; head and thorax brassy green, sometimes with a coppery tint; elytra bright cupreous, the sides brassy green, of the male shining, of the female sub-opaque. Head smooth, with an elongate shallow fovea on each side between the eyes. Thorax a trifle broader than long, slightly narrowed behind, posterior angles nearly rectangular, convex in front, depressed behind, smooth, with a deeply impressed central longitudinal line, and two deep linear rugose longitudinal impressions on each side at the base, the outer one bounded externally by a ridge. Elytra a trifle wider than the thorax, elongate, deeply striate, the striæ impunctate, interstices somewhat convex, the third with three large punctures. Beneath greenish black. Length, 5–6 lines.

Local, but widely distributed. Charlton, Weybridge, Toll-cross near Glasgow.

Fig. 30. ZABRUS OBESUS, *Dejean, Curtis.* (*F.* Feroniidæ. *G.* Zabrus, *Clairville.*)

Robust, convex, black, thorax with a brassy, elytra with a greenish, olivaceous or coppery tint. Head wide with two shallow impressions between the antennæ. Thorax transverse, slightly narrowed in front, its sides rounded, base broadly emarginate, margins punctate, disk smooth, with a faint central longitudinal impressed line. Elytra ovate, striate, the striæ smooth or very minutely and obsoletely punctate, interstices slightly convex, with a series of punctures on the lateral margin. Beneath black; tibiæ and tarsi pitchy red. Length, 7–8 lines.

Inhabits the Pyrenees, but only at high altitudes: introduced, but erroneously, into the British List on the authority of a specimen said to have been taken in Devonshire by Dr. Leach.

Fig. 31. CALLISTUS LUNATUS, *Fab., Curtis.* [*F.* Chlæniidæ. *G.* Callistus, *Bonelli.*]

Head black with a blue tint, rather thickly and coarsely punctate, smooth in front; palpi, mandibles and two basal joints of antennæ rusty-red. Thorax rusty-red, heart-shaped, convex, very thickly, somewhat less coarsely punctate than the head, posterior angles acute. Elytra ovate, faintly striate, the dorsal striæ with a few indistinct punctures, yellow, each with three bluish-black patches, the first small, transverse, at the base just above the shoulder; the second transverse, situate a little behind the middle, and extending from the lateral margin to the first stria; the third occupying the apex with the exception of its extreme tip, united with the second at the outer margin and extending to the suture. Beneath bluish-black. Thighs black, testaceous at the base; tibiæ pale yellow, with the base and tips black, tarsi pitchy-brown. Length, 3 lines.

Local, and apparently restricted to the chalk. Dover, Canterbury, and Folkestone, Kent; Mickleham, near Reigate, and near Croydon, Surrey.

Fig. 32. CHLÆNIUS SULCICOLLIS, *Payk., Curtis.* [*F.* Chlæniidæ, *G.* Chlænius, *Bonelli.*]

Above black, opaque. Head with two longitudinal impressions between the antennæ. Thorax transverse, narrowed in front, its sides reflexed, with a dense, short, yellowish-brown pubescence, punctate, the punctures coarse and sparse on the anterior portion, fine, thick, and somewhat confluent behind; central longitudinal channel broad and ruggedly punctate, dilated posteriorly, not extending to the base, and on each side a broad punctate impression ascending as far as the middle.

7

PLATE IV.—*Continued.*

Elytra finely striate, the striæ very minutely punctate, interstices thickly coriaceous, clothed with a short decumbent brown pubescence interspersed with yellow hairs, which are usually more numerous on the alternate interstices. Underside and legs shining black. Length, 6 lines.

A single example only of this beautiful species has hitherto occurred in Britain : it was found, dead, by the late Mr. Charles Curtis, at Covehithe on the Suffolk coast.

Fig. 33. BADISTER UNIPUSTULATUS, *Bonelli.* [*F.* Licinidæ. *G.* Badister, *Clairv.*] (B. cephalotes, *Dejean, Curtis.*)

Head bluish-black, smooth, nearly as wide as the thorax ; palpi reddish-yellow, the apex of the terminal joint dusky ; antennæ reddish-yellow, apex of the first and second, and the three or four following, entirely pitchy-brown. Thorax reddish-yellow, transverse, much narrowed behind, with a deep, broad, central, longitudinal channel : posterior angles rounded, a broad, deep impression on each side behind. Scutellum reddish-yellow. Elytra reddish-yellow, each with two bluish-black patches, the first transverse, situate in the middle, the second lunate, at the apex, and frequently united at the sides with the first ; finely striate, the striæ impunctate, interstices flat, smooth. Beneath black, the flancs of the mesothorax and the legs yellow. Length, 3½ lines.

Local, but widely distributed. It frequents marshy places. and has occurred, occasionally in plenty, at Shepherd's Bush, Newark, Winterbourne-stoke, Holme Fen, &c. &c.

Fig. 34. FERONIA (LAGARUS) INÆQUALIS, *Marsham.* [*F.* Feroniidæ. *G.* Feronia, *Lutr. s-G.* Lagarus, *de Chaudoir.*] (Argutor longicollis, *Curtis.*)

Oblong, depressed, reddish or pitchy brown. Head with two deep impressions between the eyes ; palpi, and antennæ red. Thorax a little longer than broad, narrowed behind, the sides rounded anteriorly, nearly straight posteriorly, smooth, posterior angles acutely rectangular, the base with a deep elongate depression on each side. Elytra oblong, sides parallel, deeply striate, the striæ with large, deep, closely set punctures, interstices flat, smooth, the third with several large punctures towards the apex. Beneath thickly punctate at the sides. Legs red. Length, 2¼-2¾ lines.

Generally distributed. Highgate and Hampstead, near London ; Gravesend, Kent ; Northfleet, Essex ; near Croydon, Surrey, &c. It frequents humid places.

Fig. 35. CALATHUS PUNCTIPENNIS, *Germar.* [*F.* Feroniidæ. *G.* Calathus, *Bonelli.*] (C. latus, *Dej., Curtis.*)

Pitchy black, with the palpi and antennæ, and occasionally the lateral margins of the thorax, ferruginous. Thorax wide behind, narrow in front, posterior angles rectangular rounded at their summits, base punctured, with two shallow elongate impressions on each side. Elytra ovate, striate, the striæ thickly punctate, interstices flat, smooth, the third and fifth with a series of punctures contiguous to the striæ. Legs brownish black. Length, 5½-6½ lines.

A native of southern Europe, and said to have been taken by Dr. Leach in Devonshire.

Fig. 36. ANCHOMENUS (AGONUM) AUSTRIACUS, *Fab.* [*F.* Feroniidæ. *G.* Anchomenus, *Erichson. s-G.* Agonum, *Bonelli.*] (Agonum austriacum, *Curtis.*)

Head and thorax bright coppery red, with a greenish tint. Elytra bright green with a silky lustre, the suture and sides frequently with a coppery red hue. Head faintly wrinkled in front, smooth posteriorly, with a shallow elongate impression on each side, between the antennæ ; parts of mouth and antennæ black, the three basal joints of the latter with a faint greenish tint. Thorax a trifle broader than long, slightly narrowed behind, the sides rounded, the lateral margins reflexed ; posterior angles rounded : a large, shallow impression on each side at the base. Elytra finely striate, the striæ finely punctate, interstices flat, smooth, the third with six punctures, of which the two anterior are usually contiguous to the third stria, the four posterior to the second stria. Beneath dark green with a faint brassy tint. Legs black, the thighs with a faint green tint. Length, 3¼-4 lines.

Widely distributed throughout Europe, but apparently excessively rare in Britain, the only recorded localities being Kingsbridge, Devonshire : Clengre. Gloucestershire ; and Cornwall.

PLATE V.

Fig. 37. AÉPUS MARINUS, *Stroem.* [*F.* Trechidæ. *G.* Aëpus, *Curtis.*] (A. fulvescens, *Curtis.*)

Oblong, depressed, testaceous yellow, smooth. Head large, eyes minute. Thorax heart-shaped, with a central longitudinal impressed line, and a shallow impression on each side behind, posterior angles rectangular, their extreme summits obtuse. Elytra oblong, sides nearly parallel, with more or less apparent punctate striæ. Underside and legs pale testaceous. Length, 1 line. Inhabits the coast, beneath stones, below high-water mark.

Fig. 38. TRECHUS MICROS, *Herbst.* [*F.* Trechidæ. *G.* Trechus, *Clairv.*] (Blemus micros, *Curtis.*)

Head reddish brown, with two deep oblique impressions on the forehead, antennæ and parts of the mouth reddish yellow. Thorax reddish brown, narrowed posteriorly, rounded at the sides in front, with a deep central longitudinal furrow, and a large impression on each side behind, posterior angles nearly rectangular. Elytra reddish yellow, each with an ill-defined dusky patch near the apex, oblong, finely striate, the striæ obliterated externally, faintly punctate, the first and third united at the base, interstices flat, thickly and minutely punctate, the fourth with two large punctures. Abdomen and legs reddish yellow, breast reddish brown. Length, 2 lines.

Widely, but sparingly distributed : inhabits the roots of tufts of coarse grass and rushes in marshy places.

Fig. 39. MASOREUS WETTERHALII, *Gyll.* [*F.* Lebiidæ. *G.* Masoreus, *Dej.*] (M. luxatus, *Curtis.*)

Reddish brown, smooth and shining. Head dusky, antennæ and parts of mouth testaceous red. Thorax transverse, emarginate in front, obliquely truncate on each side at the base, the sides rounded, with a distinct central longitudinal channel, posterior angles very obtuse. Elytra with a large ill-defined dusky patch occupying the apical two-thirds, ovate, striate, the striæ minutely punctate ; interstices flat, smooth. Legs red. Length, 2–2½ lines.

Inhabits the coast, occurring, occasionally in plenty, at Deal ; the Isle of Portland ; Sheerness, &c.

Fig. 40. CILLENUM LATERALE, *Leach, Curtis.* [*F.* Bembidiidæ. *G.* Cillenum, *Curtis.*]

Head and thorax shining, bright brassy green, with a faint coppery tinge. Head with a deep elongate impression on each side between the antennæ, which are reddish yellow, with the apex and parts of the mouth reddish brown. Thorax heart-shaped, constricted behind, convex, covered with faint transverse wrinkles, posterior angles rectangular ; base depressed, with a broad, shallow fovea on each side. Elytra sub-opaque, testaceous, with an obscure dusky cloud posteriorly, elongate, flattish, sides parallel ; striate, the striæ broad and shallow, very indistinctly punctate ; interstices convex, finely reticulate, the third wide, with four or five large punctures contiguous to the third stria. Beneath black, with a greenish tint ; margin of thorax and abdomen reddish yellow. Legs testaceous. Length, 1¾–2 lines.

Abundant near Liverpool ; on the Chesil Bank, Isle of Portland ; Smallmouth Sands, near Weymouth ; vicinity of Harwich, &c.

Fig. 41. NOTIOPHILUS RUFIPES, *Curtis.* [*F.* Elaphridæ. *G.* Notiophilus, *Dumeril.*]

Above shining brassy black. Head broad, deeply striate in front, punctate behind ; antennæ and palpi pitchy black, red at the base. Thorax heart-shaped, constricted behind, flattish, thickly and coarsely punctate, a small space on the disc smooth, with a conspicuous central longitudinal channel, and an impression on each side at the base. Elytra oblong, striate, the striæ closely and coarsely punctate, those exterior to the broad glabrous discoidal space, approximate, extending to

C   9

PLATE V.—*Continued.*

the apex, the second interstice with a large impression. Beneath black. Legs red, terminal joints of tarsi and occasionally the posterior thighs pitchy brown. Length, 2¼–3 lines.

Near Carlisle; at Shirly Common, near Croydon, Surrey; and at Darenth Wood, and Greenwich Park, Kent: beneath dead leaves in shady places.

Fig. 42. ELAPHRUS ULIGINOSUS, *Fab. Curtis.* [*F.* Elaphridæ. *G.* Elaphrus, *F.*]

Above brassy green or brown, the head, sides of thorax, and sometimes of the elytra, coppery. Head thickly, rather finely, punctate, constricted and depressed behind the eyes, a transverse ridge between the antennæ, an elongate curved impression on each side, and a shallow fovea on the forehead : mandibles, palpi, and antennæ black, three basal joints of the latter brassy green. Thorax wider than the head, its greatest width in the middle, narrowed in front, constricted behind, the sides rounded and dilated anteriorly, posterior angles acute and slightly prominent : thickly punctate ; central impressed longitudinal line terminating in front in two curved foveæ, dilated behind, reaching neither the anterior nor posterior margin, on each side, equidistant from it and the lateral margin, a round fovea ; hind margin obliquely truncate on each side, with a large impression at the angles. Elytra ovate, each with four rows of ocellated violet semi-opaque impressions united by longitudinal nearly smooth ridges ; interstices finely and rather thickly punctate. Underside and thighs brassy green; tibiæ and tarsi bluish green. Length, 3½–4 lines.

Frequents peaty bogs. Found at Rannoch ; on the Grampians ; Whittlesea Mere, &c.

Fig. 43. HALIPLUS FULVUS, *Fab.* [*F.* Haliplidæ. *G.* Haliplus, *Latr.*] (H. ferrugineus, *Curtis.*)

Rust-red, elytra with oblong fuscous patches, elongate-ovate, convex. Head minutely and sparsely punctate, palpi and antennæ testaceous. Thorax longer than broad, narrowed in front, anterior margin slightly produced in the middle, posterior margin bi-sinuate, punctate, punctures somewhat closely set in front and behind, sparser on the disc. Elytra ovate, obliquely truncate at the apex, sutural angle produced, each with ten rows of large punctures ; interstices flat, with a few small irregularly disposed punctures. Length, 2 lines.

Common in ponds and ditches throughout the country.

Fig. 44. BLETHISA MULTIPUNCTATA, *Linn., Curtis.* [*F.* Elaphridæ. *G.* Blethisa, *Bonelli.*]

Brassy brown, margins of thorax and elytra brassy green, antennæ and legs black. Head with a transverse impression between the antennæ, a deep bi-arcuate fovea on each side near the eyes, and a transverse impressed line behind them, smooth in front, finely and sparsely punctate posteriorly. Thorax much wider than the head, transverse, slightly narrowed in front, sides gently rounded, lateral margins reflexed, the sides and base ruggedly punctate, posterior angles acute, hinder margin with a large deep punctate impression on each side bounded externally by a longitudinal ridge. Elytra oblong, wider than the thorax, each with nine shallow striæ, the striæ finely remotely punctate, interstices uneven, the third with four or five, the fifth with two large impressions. Length, 5–5½ lines.

Local, but widely distributed. Conceals itself at the roots of plants growing on the muddy margins of lakes, &c.

Fig. 45. HYDROPORUS DECORATUS, *Gyll., Curtis.* [*F.* Hydroporidæ. *G.* Hydroporus, *Clairv.*]

Reddish brown, head, anterior and lateral margins of thorax, base of antennæ, legs, sides and two spots on each elytron ferruginous. Head finely and sparsely punctate. Thorax transverse, narrowed and widely emarginate in front, sides slightly rounded, base bi-sinuate, rather coarsely and somewhat thickly punctate in front and behind, very finely and sparsely on the disc. Elytra with the external margin and a sub-humeral and a sub-apical transverse spot united thereto rust-red, coarsely and sparsely punctate, the interstices with finer punctures. Length, 1¼ line.

The localities registered for this species are: York ; Norfolk ; Northampton ; and near London ; apparently scarce.

10

PLATE VI.

Fig. 46. NOTERUS SPARSUS, *Marsham, Curtis.* [*F.* Colymbetidæ. *G.* Noterus, *Clairv.*]

Head, thorax, antennæ, and legs rust-red, elytra reddish brown. Head large, convex, smooth. Thorax transverse, narrowed in front, anterior margin widely emarginate, sides slightly rounded, base sinuate, posterior angles rectangular, smooth. Elytra obconic, gradually narrowed from the base to the apex, with large scattered punctures tending to striæ at the base. Length, $2\frac{1}{4}$–$2\frac{3}{4}$ lines. Common in ponds and ditches.

Fig. 47. HYDROPORUS DAVISII, *Curtis.* [*F.* Hydroporidæ. *G.* Hydroporus, *Clairv.*]

Above pale testaceous, oblong ovate, wide, depressed. Head with an arcuate black mark between the eyes. Thorax transverse, narrowed in front, anterior margin widely and deeply emarginate, the sides slightly rounded, the base with a short deep longitudinal impressed line on each side, and a transverse black band frequently interrupted in the middle. Elytra broad ovate, each with the suture, six longitudinal lines, more or less confluent, and two lateral patches black. Underside black. Tips of palpi, antennæ, and tarsi fuscous. Length, 2–$2\frac{1}{4}$ lines.
North of England.

Fig. 48. COLYMBETES CICUR, *Fab.* [*F.* Colymbetidæ. *G.* Colymbetes, *Clairv.*]
(C. Consobrinus, *Curtis.*)

Elongate ovate, slightly convex. Head black, labrum and a large round spot on the crown reddish yellow, thickly and minutely punctate; antennæ and palpi testaceous. Thorax pitchy black, anterior margin narrowly, sides broadly, reddish yellow, and usually a longitudinal streak on the disc of the same colour, transverse, narrowed and deeply emarginate in front; sides gently rounded; base slightly sinuate; a short central impressed longitudinal line; posterior angles nearly rectangular; irregularly punctate. Scutellum and elytra reddish yellow, the latter irrorated with innumerable minute round black spots, the outer margin and four longitudinal lines on each reddish yellow, elongate ovate, each with three rows of punctures. Underside and legs reddish yellow. Length, $7\frac{1}{2}$–8 lines.
Inhabits the Cape of Good Hope.

Fig. 49. HYDATICUS CINEREUS, *Linn., Curtis.* [*F.* Dytiscidæ. *G.* Hydaticus, *Leach.*]

Ovate, convex. Head yellow, the base and a semicircular mark on the forehead black, finely and rather thickly punctate; antennæ and palpi reddish yellow. Thorax yellow, the anterior and posterior margins broadly black, transverse, narrowed and broadly emarginate in front, sides very gently rounded. Scutellum black. Elytra pitchy black, with innumerable minute round yellow spots distributed over the entire disc, external margins yellow; ovate, exceedingly minutely and rather thickly punctate, each presenting traces of three longitudinal rows of large punctures. Underside and legs reddish yellow. Length, $6\frac{1}{2}$–7 lines.
Whittlesea Mere, Hunts.

Fig. 50. DYTISCUS DIMIDIATUS, *Bergstr., Curtis, ♀.* [*F.* Dytiscidæ. *G.* Dytiscus, *Linn.*]

Elongate ovate, above olive-green, front of head and a triangular patch on the crown reddish yellow. Thorax with the lateral margins broadly, the anterior margin narrowly, reddish yellow. Elytra with the lateral margins broadly and a narrow oblique line at the apex reddish yellow; in the female each with ten longitudinal furrows extending to the middle, in the male smooth, each with three rows of punctures. Beneath testaceous yellow, the sutures dusky. Legs rust-red, coxal processes obtuse. Length, 17–18 lines.
Near Cambridge; Yaxley Fen, and Whittlesea, Hunts; Salop, &c.

11

PLATE VI.—*Continued.*

Fig. 51. GYRINUS BICOLOR, *Payk., Curtis.* [*F.* Gyrinidæ. *G.* Gyrinus, *Geoffroy.*]

Narrow, sub-parallel, convex, glossy black with a bluish tint. Head with two small round impressions between the eyes; palpi reddish yellow; antennæ fuscous. Thorax transverse, narrowed and emarginate in front, sides gently rounded, base sinuate. Elytra elongate ovate, each with ten rows of brassy punctures, interstices smooth. Beneath black, reflexed margin of elytra, sides of thorax, apex of abdomen, and legs yellowish red. Length, 3–3¼ lines.

Occurs, not uncommonly, near the coast, in salt marshes, and ditches of brackish water.

Fig. 52. ACILIUS CANALICULATUS, *Nicolai,* ♀. [*F.* Dytiscidæ. *G.* Acilius, *Leach.*]
(A. caliginosus, *Curtis.*)

Ovate, depressed. Head pitchy black, parts of the mouth, antennæ, and an arcuate patch between the eyes reddish yellow, finely reticulate and sparsely minutely punctate. Thorax pitchy black, margins, and a central transverse band dilated posteriorly at each end, and sometimes interrupted in the middle, reddish yellow; transverse, narrowed and emarginate in front, sides slightly rounded, base sinuate, posterior angles acute; finely reticulate and distinctly punctate, the punctures sparse in front and in the middle, closely set at the sides. Scutellum black, finely reticulate. Elytra reddish yellow, covered, save along the suture and at the sides, with innumerable small confluent black spots, broad ovate, dilated at the sides behind the middle, minutely punctate, each with three indistinct longitudinal ridges. Underside and legs reddish yellow, meso- and meta-thorax and base of abdominal segments frequently pitchy black. Length, 6–7 lines.

The female has the thorax depressed, especially at the sides posteriorly, and the elytra have each four wide longitudinal channels clothed with long yellowish red pubescence.

Found in the Fens of Huntingdonshire and Lincolnshire; in the north of England; and in Scotland.

Fig. 53. CYBISTER ROESELII, *Fab., Curtis,* ♀. [*F.* Dytiscidæ. *G.* Cybister, *Curtis.*]

Ovate, dilated behind, depressed. Head olive-green, shining, very minutely and sparsely punctate, labrum, epistome, palpi, and antennæ pale yellow. Thorax olive-green, lateral margins broadly, centre of the anterior and posterior margins very narrowly, yellow; transverse, narrowed and emarginate in front, sides slightly rounded, base sinuate, posterior angles acute, with an exceedingly fine and sparse punctation, and a few large punctures at the sides and along the anterior margin. Elytra ovate, brownish olive-green, with a broad pale yellow sub-marginal band, and a narrow oblique stripe at the apex, each with three rows of punctures. Underside and legs reddish yellow. Length, 12–15 lines.

The female differs from the male in having the thorax covered with deeply impressed curved lines, and the elytra furnished to within about a fifth of their length from the apex with closely set waved longitudinal strigæ.

The only example hitherto detected in Britain "was found the 30th September, 1826, in a puddle at Walton, Essex."

Fig. 54. PARNUS PROLEFERICORNIS, *Fab.* [*F.* Parnidæ. *G.* Parnus, *Fab.*] (P. impressus, *Curtis.*)

Fuscous, clothed with a dense short yellowish-grey pubescence, elongate, sub-cylindrical, finely and rather thickly punctate. Head depressed in front; antennæ approximate, black, the club reddish yellow. Thorax a little narrower than the elytra, narrowed in front, with a slightly curved impressed line on each side, and occasionally two foveæ near the base. Elytra with feeble traces of striæ. Length, 2¼–2¾ lines.

Common on the muddy margins of ponds and ditches.

12

PLATE VII.

Fig. 55. HYDRÆNA TESTACEA, *Curtis.* [*F.* Elophoridæ. *G.* Hydræna, *Kugelann.*]

Head black, minutely and thickly punctate; antennæ and palpi rufo-testaceous. club of the former fuscous. Thorax pitchy black or brown, with a narrow rufous band on the posterior, and a wider one on the anterior margin, elongate, narrowed behind ; the sides rounded in front, sub-angular a little behind the middle ; coarsely and confluently punctate, with an indistinct elongate fovea on each side in front, and a faint impressed transverse line near the anterior and posterior margins. Elytra reddish brown, external margin paler, elongate ovate, rounded at the apex, closely striate-punctate, the punctures large and confluent. Underside dull black. Legs reddish yellow. Length, ½ line.

Widely distributed, amongst aquatic plants in stagnant water, and occasionally beneath stones on the margins of sluggish streams.

Fig. 56. ELOPHORUS RUGOSUS, *Olivier.* [*F.* Elophoridæ. *G.* Elophorus, *F.*]
(E. fennicus, *Curtis.*)

Pale ferruginous, head and thorax darker. Head short, deflexed, coarsely granulate ; antennæ and palpi rufo-testaceous. Thorax transverse, slightly narrowed in front, sides rounded, posterior angles rectangular, granulate, with four irregular interrupted longitudinal ridges. Elytra tessel-lated with black, a trifle wider at the base than the thorax, slightly dilated posteriorly, each with ten rows of punctures, the alternate interstices forming elevated ridges. Tibiæ denticulate externally. Length, 2–2¾ lines.

Found, but not abundantly, throughout Britain.

Fig. 57. HYDROCHUS ELONGATUS, *F.*, *Curtis.* [*F.* Elophoridæ. *G.* Hydrochus, *Leach.*]

Elongate, black with a green or brassy tint. Head thickly and coarsely punctate, forehead with three longitudinal foveæ ; antennæ and palpi ferruginous. Thorax elongate, narrowed behind, sides nearly straight, less thickly and a trifle more coarsely punctate than the head, with five foveæ, of which the three anterior are disposed transversely a little before the middle, the two posterior, sometimes confluent, close to the posterior margin. Elytra elongate, each with ten rows of large punctures, the 1st (sutural), 3rd, 4th, 5th, 7th, and 9th interstices elevated into ridges, the 1st and 3rd throughout their entire length, the 4th at the apical two-thirds, the others at the basal two-thirds. Legs reddish brown, thighs darker. Length, 2 lines.
Not rare.

Fig. 58. ELMIS VOLKMARI, *Panz., Curtis.* [*F.* Elmidæ. *G.* Elmis, *Latr.*]

Elongate, black with a faint brassy tint, slightly shining ; antennæ ferruginous, dusky at the tips. Thorax a little longer than broad, narrowed in front, somewhat thickly and finely punctate, with an impressed longitudinal line on each side. Elytra punctate-striate, interstices flat, very finely punctate. Beneath black, apex of abdomen usually pitchy red. Legs reddish brown, tarsi rufous. Length, 1½–1¾ lines.
Beneath stones in streams.

Fig. 59. HYDROPHILUS PICEUS, *L.* [*F.* Hydrophilidæ. *G.* Hydrophilus, *Geoff.*]
(Hydröus piceus, *Curtis.*)

Elongate ovate, above moderately convex, black, smooth, shining. Head smooth, with an oblique sparsely punctate impression on each side between the antennæ, and a thickly punctate one near each eye ; antennæ and palpi pitchy red, club of the former fuscous. Thorax transverse, narrowed in front. sides rounded. Elytra ovate, each with eight rather faint impunctate longitud-inal striæ, interstices flat, the second with a somewhat irregular single row of punctures, the fourth

13

PLATE VII. — *Continued.*

and sixth with a very irregular double one. Beneath black, meso- and meta-thorax thickly clothed with golden pubescence; sternum with a furrow in front, and an impressed line behind; abdomen keeled in the centre throughout its whole length. Length, 16–20 lines.
Common, in ponds, throughout England.

Fig. 60. OCHTHEBIUS PUNCTATUS, *Steph.* [*F.* Elophoridæ. *G.* Ochthebius, *Leach.*] (O. hibernicus, *Curtis.*)

Elongate ovate, convex, black, with a brassy or coppery tint. Head minutely and rather thickly punctate, with two deep foveæ in front; antennæ and palpi reddish yellow, club of the former fuscous. Thorax transverse, narrowed in front and behind, sides rounded, minutely and thickly punctate, with a central longitudinal impressed line, and on each side of it a shallow interrupted oblong fovea, posterior angles with a large shallow impression. Elytra oblong, with an abbreviated sutural stria, irregularly rugulose-punctate. Beneath ferruginous, with a brassy or coppery tint, pubescent. Legs reddish yellow. Length, 1¼ line.
Rare: near London; Bristol, &c.; and in Ireland.

Fig. 61. OCHTHEBIUS EXSCULPTUS, *Müller.* (Female.) [*F.* Elophoridæ. *G.* Ochthebius, *Leach.*] (Enicocerus Gibsoni, *Curtis.*)

Ovate, above black, with a brassy or coppery hue. Head finely and sparsely punctate, forehead with two oblique foveæ united posteriorly. Thorax heart-shaped, punctate, in the male (E. viridiæneus, *Steph., Curtis*) black, with the lateral margins brassy, a central longitudinal channel, and on each side of it an oblong fovea, and an oblique lateral impression near the base; in the female, brassy throughout, with four foveæ arranged in a quadrangle, two on each side of the central channel, and a large ovate oblique lateral impression behind. Elytra broadly ovate, each with ten punctate striæ, the punctures large and closely set; interstices narrow, convex, smooth, the 3rd, 5th, and 7th more salient than the others. Beneath black, with a silvery pubescence. Legs yellow, apex of femora and tarsi fuscous. Length, ¾–1 line.
Not uncommon in the north of England and in Scotland, beneath stones in streams.

Fig. 62. HETEROCERUS OBSOLETUS, *Curtis.* [*F.* Heteroceridæ. *G.* Heterocerus, *Bosc.*]

Oblong, rather convex, pitchy black, with a dense erect brown pubescence. Head thickly punctate, mandibles pitchy red. Thorax transverse, thickly punctate, with a ferruginous spot at the anterior angles. Elytra finely and thickly punctate, each with a ferruginous spot at the base, three placed transversely a little before the middle, three a little behind the middle, and two near the apex, one or more of which are sometimes very obscure or entirely absent. Legs black, with a brownish pubescence, tarsi reddish brown. Length, 2–3 lines.
Common on the muddy margins of ditches and pools of brackish water on the east and south coasts.

Fig. 63. HYDROUS CARABOIDES, *L.* [*F.* Hydrophilidæ. *G.* Hydrous, *Brullé.*] (Hydrophilus caraboides, *Curtis.*)

Ovate, convex, black, above shining. Head with a shallow, oblique, very coarsely and sparsely punctate impression on each side in front: antennæ and palpi reddish yellow, the club of the former black. Thorax transverse, narrowed in front, slightly rounded at the sides, with a few large dispersed punctures near the lateral margins. Elytra dilated behind the middle, indistinctly striate, interstices smooth, the 3rd, 7th, 9th, and 10th with an irregular row of large punctures. Beneath opaque, with a dense short yellowish brown pubescence, terminal segment of abdomen with a smooth shining space at the apex. Legs pitchy black. Length, 8–10 lines.
Common in ponds and ditches.

14

PLATE VIII.

Fig. 64. ONTHOPHAGUS TAURUS, *L.*, *Curtis.* (Male.) [*F.* Copridæ. *G.* Ontho-phagus, *Latr.*]

Black, thorax with a greenish, brassy, or coppery hue. Head elongate, somewhat produced in front, furnished in the male with two long curved horns, in the female with two transverse ridges. Thorax transverse, narrowed in front. sides rounded posteriorly, convex, finely and sparsely punctate. Elytra rather flat, dull black or brown, each with nine faint punctate striæ, interstices minutely and very sparsely punctate. Legs black, tarsi reddish brown, anterior tibiæ quadri-dentate externally. Length, 4-5 lines.
Very rare. New Forest, Hampshire, in dung.

Fig. 65. SPERCHEUS EMARGINATUS, *Schaller*, *Curtis.* [*F.* Spercheidæ. *G.* Sper-cheus, *Kugelann.*] *

Ovate, convex, yellowish brown, head and disc of thorax fuscous. Head emarginate in front, thickly and finely punctate, forehead with an interrupted arcuate impression. Thorax transverse, narrowed in front, rounded at the sides, thickly and rather coarsely punctate. Elytra broadly ovate, somewhat thickly and coarsely punctate, with from six to eight indistinct smooth longitudinal ridges. Beneath fuscous. Legs reddish yellow, sometimes pitchy red, tarsi fuscous at the tip. Length, 3-3½ lines.
Very rare. Found at the roots of aquatic plants, in stagnant water.

Fig. 66. GEOTRUPES VERNALIS, *L.* [*F.* Geotrupidæ. *G.* Geotrupes, *Latr.*] (G. lævis, *Curtis.*)

Broad-ovate, convex, black, above with a bright blue, violet, or green tint, beneath blue. Head rugulose in front, with a short longitudinal ridge. Thorax transverse, narrowed in front, rounded at the sides, thickly and finely punctate, with larger scattered punctures. Scutellum finely punctate at the base. Elytra very obsoletely striate-punctate, interstices smooth. Length, 5-9 lines.
Not uncommon.

Fig. 67. ODONTÆUS MOBILICORNIS, *Fab.* (Male.) [*F.* Geotrupidæ. *G.* Odontæus, *Klug.*] (Bolbocerus mobilicornis, *Curtis.*)

Ovate, above testaceous, or reddish brown, or black, paler beneath. Head ruggedly punctate, with a short raised longitudinal line in front, in the male with a long slender curved moveable horn, in the female with two obsolete tubercles; antennæ reddish yellow. Thorax transverse, narrowed in front, coarsely punctate, with a broad central longitudinal channel, in the male excavated in front with a broad flat horn on each side a little within and behind the anterior angles, and an acute tubercle in front on either side of the central channel. Scutellum finely and sparsely punctate, with several large deep punctures at the base. Elytra deeply striate, the striæ strongly and somewhat remotely punctate; interstices smooth, convex. Beneath ferruginous, with a sparse long yellow pubescence. Legs pitchy red; tibiæ brown, the anterior with from six to eight denticulations on the outer edge. Length, 3½-4 lines.
Very rare, but apparently distributed throughout the southern counties. It inhabits dung, and flies towards sunset.

* In Illiger's *Verzeichniss der Käfer Preussens*, 211 (1798).

15

PLATE VIII.— *Continued.*

Fig. 68. TROX SABULOSUS, *L., Curtis.* [*F.* Trogidæ. *G.* Trox, *Fab.*]

Oblong, black, sub-opaque. Head thickly and coarsely punctate; antennæ reddish brown. Thorax transverse, slightly narrowed in front, gently rounded at the sides; posterior angles rectangular, prominent; thickly and coarsely punctate, with a broad central longitudinal impression, and on each side, on the disc, a large oblong, frequently interrupted, fovea. Elytra faintly striate, the striæ coarsely punctate, the alternate interstices raised and furnished with a row of fascicles of yellowish grey bristles. Length, 4–1½ lines.

Found, but not commonly, throughout the country.

Fig. 69. SPHÆRIDIUM BIPUSTULATUM, *F.* [*F.* Sphæridiidæ. *G.* Sphæridium, *Fab.*] (S. quadrimaculatum, *Curtis.*)

Short ovate, above shining black. Head thickly and very finely punctate, antennæ reddish brown. Thorax with the lateral margins narrowly reddish yellow; transverse, narrowed in front, rounded at the sides, posterior angles acute; very finely and thickly punctate. Elytra with the outer margin and a lunate patch at the tip reddish yellow, and generally a large blood-red humeral spot. Beneath dull black. Legs reddish yellow, thighs with a transverse black patch. Length, 1¾–2½ lines.

Common in cow-dung.

Fig. 70. APHODIUS VILLOSUS, *Gyll., Curtis.* [*F.* Aphodiidæ. *G.* Aphodius, *Illiger.*]

Elongate, brown, clothed with long yellowish grey pubescence. Head thickly punctate. Thorax transverse, coarsely, deeply and rather sparsely punctate, posterior margin sinuate. Scutellum elongate, triangular, faintly punctate. Elytra oblong, each with seven broad shallow striæ, the striæ remotely punctate; interstices convex. Legs reddish brown. Length, 1½–2 lines.

Very rare. Newmarket Heath, and in Cornwall.

Fig. 71. COPRIS LUNARIS, *L., Curtis.* (Male.) [*F.* Copridæ. *G.* Copris, *Geoffr.*]

Above shining black. Head semicircular, notched in front, of the male with an erect, more or less elongate pointed horn, of the female with a short erect horn, notched at the tip. Thorax transverse, convex, retuse in front, ruggedly punctate at the sides, with a central impressed longitudinal line. Elytra convex, each with nine shallow striæ; interstices flattish, sparsely and very finely punctate. Beneath with long ferruginous hairs. Length, 7–10 lines.

Common in dung in sandy places in the south.

Fig. 72. PSAMMODIUS SULCICOLLIS, *Illiger, Curtis.* [*F.* Asphodiidæ. *G.* Psammodius, *Gyll.*]

Oblong ovate, convex, brown. Head covered with tubercles, with a semicircular impression on the crown: clypeus emarginate in front: antennæ and palpi reddish brown. Thorax transverse, slightly narrowed in front, sides rounded, with five deep, broad, coarsely punctate, transverse, furrows, the spaces between them convex, smooth: central longitudinal channel obsolete in front, deep behind. Elytra convex, a little dilated behind the middle, each with nine broad, deep, coarsely punctate striæ; interstices convex, smooth. Underside and legs reddish brown. Length, 1½–1¾ lines.

Not uncommon on the Welsh and Lancashire coasts.

16

PLATE IX.

Fig. 73. SINODENDRON CYLINDRICUM, *L. Curtis* (Male). [*F.* Lucanidæ.
G. Sinodendron, *F.*]

Elongate, semicylindrical, black. Head coarsely and sparsely punctate, armed in front with a more or less elongate curved horn, fringed on each side near the apex with golden yellow hairs. Thorax scarcely longer than broad, slightly narrowed in front, anterior margin semicircularly emarginate, sides nearly straight ; semicylindrical, sub-perpendicularly truncate in front, the upper margin of the truncation with a prominent tooth in the centre, and an obtuse one on each side; very coarsely and sparsely punctate, with a smooth discoidal space. Elytra nearly as wide as the thorax, each with ten more or less interrupted longitudinal striæ, sparsely punctate, the punctures large, irregularly disposed, and frequently confluent.

In the female the head is tuberculated, and the thorax presents two shallow approximate foveæ in front, and both these segments are coarsely and confluently punctate. Length, 4¼-6½ lines.

Inhabits dead trees, especially ash, and occurs not uncommonly throughout the country.

Fig. 74. GNORIMUS VARIABILIS, *L.* (Male.) [*F.* Cetoniidæ. *G.* Gnorimus, *Serville.*]
(Trichius variabilis, *Curtis.*)

Ovate, depressed, black. Head emarginate in front, thickly and confluently punctate. Thorax wide behind, narrowed in front, sides rounded, thickly and coarsely punctate, with a conspicuous central longitudinal channel. Elytra much wider than the thorax, thickly and ruggedly punctate, presenting faint traces of longitudinal striæ, each with four or five small yellowish white spots, two or three on the disc, and two near the lateral margin. Terminal segment of abdomen with two large yellowish white spots on each side. Beneath black, thorax clothed with long yellowish grey pubescence; abdomen with a central longitudinal depression. Intermediate tibiæ bent in a semicircle at the base, gradually dilated from the middle to the apex ; posterior tarsi conspicuously longer than their tibiæ.

In the female the abdomen is not furrowed beneath, the intermediate tibiæ are nearly straight, and the posterior tarsi do not exceed their tibiæ in length. Length, 9-11 lines.

Rare. In old oaks, Windsor Forest, Tooting Common, near Croydon, and near Highgate.

Fig. 75. OXYTHYREA STICTICA, *L.* [*F.* Cetoniidæ. *G.* Oxythyrea, *Mulsant.*]
(Cetonia stictica, *Curtis.*)

Oblong, black with a greenish or coppery tint, sparingly clad with a long erect grey pubescence. Head coarsely punctate, deeply notched in front. Thorax wide behind, narrowed anteriorly, sides rounded in front, nearly straight behind, thickly and coarsely punctate, with a faint central longitudinal impression, and four rows of three white spots. Elytra with numerous white spots, sub-quadrate, depressed along the suture ; each with two faint longitudinal ridges, and six double rows of punctures ; interstices sparsely punctate. Beneath shining black. Anterior tibiæ with two teeth on the outer edge. Length, 4-5½ lines.

Very rare. Windsor Forest.

Fig. 76. POLYPHYLLA FULLO, *L.* (Male.) [*F.* Melolonthidæ. *G.* Polyphylla,
Harris.] (Melolontha fullo, *Curtis.*)

Oblong, above convex, pitchy black or brown, spotted and irrorated with white scales. Head thickly and finely punctate, forehead with dispersed white scales in the centre, and a dense band on each side next the eyes ; antennæ pitchy red, the club consisting of seven elongate leaflets. Thorax transverse, narrowed in front, sides rounded, with a conspicuous central furrow, and three longitudinal, more or less interrupted lines of white scales. Scutellum densely clothed with white scales divided by a dark central line. Elytra ruggedly punctate, irrorated with spots and patches of white scales. Beneath with long yellow hairs, abdomen with a short depressed grey pubescence. Anterior tibiæ bi-dentate externally near the apex.

PLATE IX.—*Continued.*

In the female the club of the antennæ is composed of six diminutive leaflets, and the anterior tibiæ are armed externally with three acute teeth. Length, 12–15 lines.

Very rare. Inhabits sandy situations, and is said to have occurred at Deal, Dover, Margate, Sandwich, and Hythe.

Fig. 77. LUCANUS CERVUS, *L., Curtis.* (Male.) [*F.* Lucanidæ. *G.* Lucanus, *Linn.* STAG BEETLES.]

Elongate, black, sub-opaque mandibles and elytra castaneous. Head much wider than the thorax, very thickly, minutely, and ruggedly punctate, its margins raised ; mandibles very long, curved, more or less distinctly serrated within, with a large tooth near the centre, and the apex bifurcate. Thorax transverse, narrowed in front, posterior angles obliquely truncate ; with a very obscure, central, longitudinal, impressed line ; thickly, minutely, and ruggedly punctate. Elytra fully thrice as long as the thorax, humeral angles dentate, very thickly, minutely, and confluently punctate. Beneath black, with a short, grey pubescence, thickly and minutely punctate. Legs elongate, black.

In the female the head is very much narrower than the thorax, its margins are not raised, and the mandibles are small, conspicuously shorter than the head, and the legs are shorter and more robust. Length (exclusive of mandibles), 14–24 lines.

Common in many places in the south of England. The larvæ reside in old trees, especially oaks, and are, it is said, from five to six years in attaining their full growth.

Fig. 78. PHYLLOPERTHA HORTICOLA, *L., var.* (*F.* Rutelidæ. *G.* Phyllopertha, *Steph.*] (Anisoplia suturalis, *Newman, Curtis.*)

Oblong, black, with a bright metallic blue or green tint, pubescent, elytra testaceous yellow. Head truncate in front, thickly, minutely, and ruggedly punctate ; antennæ reddish yellow or pitchy, club black. Thorax transverse, narrower than the elytra, narrowed in front ; sides rounded anteriorly, subsinuate posteriorly ; base bi-sinuate, posterior angles prominent ; rather minutely and somewhat thickly punctate, with a very indistinct central longitudinal furrow. Elytra slightly dilated posteriorly, varying in colour from testaceous yellow to pitchy brown, coarsely, irregularly punctate-striate ; interstices convex, smooth. Legs black, with a metallic green or blue tint. Length, 3–5 lines.

Common throughout the country.

Fig. 79. HISTER QUADRIMACULATUS, *L., Curtis.* [*F.* Histeridæ. *G.* Hister, *L.*]

Elongate, sub-quadrate, depressed, shining black. Antennæ pitchy ; club red, pubescent. Thorax transverse, narrowed in front, the sides slightly rounded, inner lateral stria entire, outer one reduced to a short line at the anterior angles. Elytra a little longer than the thorax, with a large lunate spot frequently obsolete or wanting: with a sub-humeral and three entire dorsal striæ, and occasionally indistinct traces at the apex of one or two others ; reflexed margin bi-sulcate. Propygidium with a shallow impression on each side, its margins coarsely punctate : pygidium coarsely and thickly punctate throughout. Anterior tibiæ with three stout denticulations externally. Length, 3¼–4½ lines.

Not uncommon in many places on the south coast.

Fig. 80. DENDROPHILUS PYGMÆUS, *L.* [*F.* Histeridæ. *G.* Dendrophilus, *Leach.*] (D. Sheppardi, *Curtis.*)

Broad-ovate, convex, pitchy black, slightly shining, legs and antennæ reddish brown. Head very much deflexed. Thorax transverse, narrowed in front, rounded at the sides, convex. Elytra broad, dilated a little before the middle, each with an entire marginal stria, and six fine dorsal striæ which do not reach the apex, interstices flat but slightly raised along the striæ, reflexed margins with an abbreviated furrow. All the tibiæ dilated, the anterior with five or six minute denticulations externally. Length, 1½–1¾ lines.

Inhabits the nests of the wood-ant (Formica rufa).

18

PLATE X.

Fig. 81. PLATYCERUS CARABOIDES, *L.*, *Curtis.* [*F.* Lucanidæ. *G.* Platycerus, *Geoffr.*]

Elongate, bright blue, sometimes green or violet. Head thickly and ruggedly punctate. Thorax transverse, subquadrate, more or less narrowed in front, sides rounded, finely and rather sparsely punctate, with a faint central longitudinal furrow. Elytra thrice as long as the thorax, sides parallel ; ruggedly punctate, the punctures here and there assuming irregular striæ. Beneath punctate, abdomen sparsely pubescent. Length, 5–6 lines.

Said to have been taken near Bristol ; at Oxford, and near Aberdeen.

Fig. 82. IPS QUADRIPUNCTATA, *Hbst.*, *Curtis.* [*F.* Nitidulidæ. *G.* Ips, *Fab.*]

Oblong, convex, shining black. Head triangular, rather thickly and coarsely punctate ; antennæ pitchy. Thorax transverse, narrowed in front, the sides slightly rounded, lateral margins reflexed, coarsely and somewhat thickly punctate. Elytra oblong, each with two large yellow spots, one oblique at the base, the other transverse near the apex ; somewhat finely and sparsely punctate, with three or four exceedingly fine longitudinal striæ on the disc. Legs pitchy. Length, $2\frac{3}{4}$–3 lines.

Found, but not commonly, beneath the bark of trees, especially oaks.

Fig. 83. MICROPEPLUS TESSERULA, *Curtis.* [*F.* Micropeplidæ. *G.* Micropeplus, *Latr.*]

Pitchy black, slightly shining. Head depressed in front, the base with three furrows ; antennæ pitchy red at the base. Thorax transverse, narrowed in front, sides rounded anteriorly, lateral margins flat, with a broad central longitudinal channel, and an oblong fovea on each side. Elytra conspicuously longer than the thorax, each having the suture elevated, and three longitudinal raised lines ; interstices exceedingly minutely punctate. Abdomen with five segments exposed, the three first with three longitudinal ridges. Legs pitchy red. Length, $\frac{3}{4}$ line.

Very rare. Belfast, and near Nottingham.

Fig. 84. HYDROBIUS FUSCIPES, *L.*, *var.* [*F.* Hydrophilidæ. *G.* Hydrobius, *Leach.*] (H. chalconotus, *Curtis.*)

Oblong-ovate, convex, black with a brassy hue. Head with a transverse row of large punctures in front, frequently interrupted in the middle, and with an oblique punctate impression on each side near the eyes ; antennæ and palpi pale red, the latter fuscous at the apex. Thorax transverse, narrowed in front, rounded at the sides, thickly and finely punctate, with a few large, irregularly scattered punctures near the lateral margins. Elytra ovate, thickly and finely punctate ; each with eleven striæ, which are shallow at the base, deepen gradually to the apex, and are profoundly and closely punctate ; interstices flat, the 3rd, 5th, 7th, 9th, and 11th, with large irregularly dispersed punctures. Beneath brown, pubescent, ruggedly punctate ; abdominal segments with a transverse impression at each side. Legs pitchy brown, tarsi paler. Length, $2\frac{3}{4}$–$3\frac{1}{4}$ lines.

Common in ponds and ditches.

Fig. 85. BEROSUS SIGNATICOLLIS, *Charp.** [*F.* Hydrophilidæ. *G.* Berosus, *Leach.*] (B. æriceps, *Curtis.*†)

Broad ovate, convex. Head brassy green with a coppery tint, thickly and finely punctate ; antennæ and palpi pale yellow. Thorax transverse, narrowed in front, sides nearly straight,

---

* *Horæ Entomologicæ*, 201 (1825). † *British Entomology*, v. 240 (1828).

PLATE X.—*Continued*.

posterior angles obtuse ; rather thickly and finely punctate ; pale yellow, with an ovate greenish black spot on the disc divided by a slightly raised, smooth, yellow, central, longitudinal line. Scutellum brassy black, confluently punctate. Elytra broad ovate, convex, pale yellow with several obscure black spots, punctate striate ; interstices wide, flat, each with an irregular row of large dark punctures. Beneath dull black. Legs pale yellow, base of thighs pitchy brown. Length, 2¼–2½ lines.
Common in the south.

Fig. 86. OMOSITA COLON, *L.* [*F.* Nitidulidæ. *G.* Omosita, *Eric.*] (Nitidula colon, *Curtis.*)

Oblong, black, with a short golden yellow pubescence. Head depressed in front, thickly and rather finely punctate ; antennæ ferruginous, club dusky. Thorax transverse, narrowed in front, rounded at the sides, thickly and rather finely punctate, with two approximate foveæ on the disc posteriorly, the lateral and anterior margins ferruginous. Elytra thickly and finely punctate, each with four or five patches, and the apex ferruginous. Legs reddish brown. Length, 1–1¼ lines.
Common in bones.

Fig. 87. CRYPTARCHA IMPERIALIS, *F.* [*F.* Nitidulidæ. *G.* Cryptarcha, *Shuck.*] (Strongylus imperialis, *Curtis.*)

Ovate, ferruginous, pubescent. Head thickly and finely punctate, crown dusky ; antennæ reddish brown, club fuscous. Thorax transverse, narrowed in front, rounded at the sides, posterior margin waved, thickly and finely punctate, disc dark brown. Elytra black in the centre, variegated with testaceous markings, minutely and rather thickly punctate. Beneath reddish brown, metathorax fuscous. Legs reddish yellow. Length, 1¼–1½ lines.
Not common, although widely distributed. Beneath bark of oak.

Fig. 88. ONTHOPHILUS SULCATUS, *Fab., Curtis.* [*F.* Histeridæ. *G.* Onthophilus, *Leach.*]

Broad ovate, black, shining. Head thickly punctate, with an impression in front ; antennæ reddish brown. Thorax transverse, narrowed in front, rounded at the sides, rather sparsely punctate, with five longitudinal ridges, the central one forked in front, furrowed behind, the others arcuate, and abbreviated anteriorly. Elytra with the suture raised, each with three elevated longitudinal lines, interstices finely striate, and with three rows of punctures. Legs reddish brown. Length, 1¼–1¾ lines.
Very rare. The following localities have been recorded : Norfolk; Coombe Wood, Surrey; Nottinghamshire, and the West of England.

Fig. 89. THYMALUS LIMBATUS, *Fab., Curtis.* [*F.* Trogositidæ. *G.* Thymalus, *Latr.*]

Ovate, convex, reddish brown with a brassy tint, clothed with a short erect grey pubescence. Head buried to the eyes in the thorax. Thorax transverse, sides broadly reflexed deep red ; finely and thickly punctate. Elytra with the outer margin deep red, deeply and coarsely punctate, the punctures arranged in rows near the suture, irregular on the disc and at the sides, interstices very finely and sparsely punctate. Length, 2¼–3 lines.
Found copiously beneath bark in the New Forest; near Cambridge; Sherwood Forest; the Black Forest, Perthshire, &c.

PLATE XI.

Fig. 90. BYTURUS TOMENTOSUS, *Fab.*, *Curtis*. [*F.* Telmatophilidæ. *G.* Byturus, *Latr.*]

Elongate, black, with a dense grey pubescence, antennæ and legs reddish yellow, or reddish brown with a yellowish grey pubescence, and the antennæ and legs pale yellow. Labrum obsolete; eyes small, prominent. Length, 1¾ lines.

Common throughout the spring, summer, and autumn in flowers, especially of the white-thorn, raspberry, and bramble.

Fig. 91. TRIPLAX ÆNEA, *Payk.*, *Curtis*. [*F.* Erotylidæ. *G.* Triplax, *Payk.*]

Oblong ovate, convex. Head small, triangular, rounded in front, reddish yellow, finely and sparsely punctate; antennæ pitchy black. Thorax transverse, narrowed in front, sides slightly rounded; base bi-sinuate, margined; reddish yellow, rather finely and very sparsely punctate. Scutellum reddish yellow. Elytra oblong, greenish blue, finely punctate-striate; interstices flat, exceedingly minutely and sparsely punctate. Underside and legs reddish yellow. Length, 1¾–2¼ lines.

Rare. Coombe Wood, Surrey; New Forest; Ockbrook, Derbyshire, &c.

Fig. 92. ANTHEROPHAGUS PALLENS, *Oliv.* (Female.) [*F.* Cryptophagidæ. *G.* Antherophagus, *Latr.*] (A. similis, *Curtis.*)

Oblong, red brown, with a short, grey pubescence. Head triangular, finely and rather thickly punctate. Thorax transverse, quadrate, slightly narrowed in front, sides nearly straight, finely and thickly punctate. Elytra oblong, exceedingly finely and somewhat closely punctate, each with six or seven indistinct striæ. Legs reddish yellow; anterior tibiæ robust, dilated towards the apex, outer angle rectangular.

In the male the antennæ are thick, moniliform, pitchy black, the basal and apical joints reddish brown; the legs are shorter and stouter, and the tibiæ are pitchy black at the base. Length, 1½–1¾ lines

In flowers, not common, although widely distributed.

Fig. 93. MYCETÆA HIRTA, *Marsh.*, *Curtis*. [*F.* Endomychidæ. *G.* Mycetæa, *Steph.*]

Oblong ovate, convex, reddish yellow, or red brown, shining, sparingly clothed with long, coarse, sub-erect, golden yellow hairs. Head triangular, thickly and very finely punctate; antennæ pale reddish yellow. Thorax transverse, narrowed in front, rounded at the sides, posterior angles nearly rectangular, base with a transverse impressed line, a trifle finely and more sparsely punctate than the head; on each side within the lateral margin, and somewhat remote from it, a shallow, longitudinal, impressed line, bounded outwardly by an acute ridge. Scutellum short, very broad, its apex truncate. Elytra elongate ovate, acuminate behind; striate-punctate, the punctures large, rather shallow, and somewhat remote; interstices smooth, the 1st, 3rd, 5th, and 7th, with a row of long, sub-erect, golden yellow setæ. Legs pale reddish yellow. Length, ⅔ line.

Common in cellars, boleti, and hay-stacks.

Fig. 94. CRYPTOPHAGUS POPULI, *Payk.*, *Curtis*. [*F.* Cryptophagidæ. *G.* Cryptophagus, *Hbst.*]

Oblong ovate, ferruginous red, pubescent, elytra with an elongate black patch, frequently invading the entire disc. Head triangular, thickly, deeply, and coarsely punctate. Thorax

PLATE XI.—*Continued.*

transverse, convex, slightly narrowed behind, sides with a recurved denticulation a little behind the anterior angles, and a minute tooth just before the middle, punctate like the head. Elytra oblong, rather finely and somewhat thickly punctate. Legs reddish yellow. Length, 1½–2 lines. Rare. Near Norwich; Ripley, Surrey; Gravesend, Kent.

Fig. 95. TRITOMA BIPUSTULATA, *F., Curtis.* [*F.* Erotylidæ. *G.* Triplax, *Fab.*]

Short ovate, black, shining, elytra with a large triangular, blood-red, humeral patch. Head triangular, finely and rather closely punctate; antennæ red, the basal joint and the club pitchy black. Thorax transverse, narrowed in front, sides nearly straight, base bi-sinnate, with a very short, longitudinal, impressed line in front of the scutellum; finely and rather thickly punctate. Scutellum semicircular, smooth. Elytra ovate, convex, each with eight rows of fine closely-set punctures; interstices exceedingly finely and sparsely punctate; black, with a large, blood-red, triangular spot on the shoulder, humeral callus rather prominent, pitchy black. Beneath punctate. Legs pitchy black, tarsi ferruginous. Length, 2–2½ lines.
Not uncommon in boleti, and in decaying stumps of trees.

Fig. 96. TETRATOMA ANCORA, *F., Curtis.* [*F.* Tetratomidæ. *G.* Tetratoma, *Hbst.*]

Oblong ovate, reddish yellow, elytra pitchy brown, with the outer margins and five spots reddish yellow. Head triangular, thickly and coarsely punctate. Thorax transverse, convex, narrowed in front, rounded at the sides; posterior margin waved, with a triangular impression on each side midway between the angle and the centre; coarsely and rather sparsely punctate. Elytra oblong, as coarsely but a trifle more sparsely punctate than the thorax, pitchy brown, the external margins, an elongate patch descending obliquely from the shoulder, another ascending obliquely from the outer margin near the apex, and a common ovate spot on the suture reddish yellow; these markings subject to considerable variation. Length, 1½ lines.
Very rare. Found in moss and decayed branches of oak and hornbeam, at Highgate and Colney Hatch, near London; and in the north of England.

Fig. 97. MYCETOPHAGUS PICEUS, *Fab., Curtis.* [*F.* Mycetophagidæ. *G.* Mycetophagus, *Hellwig.*]

Ovate, rather convex, pitchy black, pubescent. Head ferruginous in front, thickly and finely punctate, antennæ ferruginous at the base and apex, dusky in the middle. Thorax transverse, narrowed in front, rounded at the sides: posterior margin bi-sinuate, with an ovate impression at each side: reddish brown, thickly and finely punctate. Elytra oblong, conspicuously punctate-striate at the base, obsoletely so at the apex; interstices flat, exceedingly finely punctate; with a broad, interrupted band at the base, a patch on the outer margin, a narrow, interrupted, transverse band behind the middle, and an ovate spot within the apex reddish yellow. Beneath reddish brown. Legs reddish yellow. Length, 2¼–2½ lines.
The markings on the elytra vary considerably, and it is difficult to find two individuals precisely alike in this respect.
Not uncommon in fungi on decayed oaks.

Fig. 98. TYPHLEA FUMATA, *L., Curtis.* [*F.* Mycetophagidæ. *G.* Typhæa, *Curtis.*]

Oblong ovate, depressed, rufo-testaceous, thickly pubescent. Head triangular, thickly and finely punctate, with an arcuate impression between the antennæ. Thorax transverse, narrowed in front, sides rounded; posterior margin bi-sinuate, with a broad, shallow impression on each side; thickly and finely punctate. Elytra oblong, punctate-striate; interstices thickly and finely rugulose-punctate, each with a row of long, decumbent, golden yellow hairs. Length, 1–1½ lines.
Common in dung-heaps, hay-stack refuse, &c.

PLATE XII.

Fig. 99. CATOPS MORIO, *Fab.* [*F.* Silphidæ. *G.* Catops, *Fab.*] (C. dissimulator, *Curtis.*)

Oblong ovate, convex, black, pubescent, sub-opaque. Head finely and thickly punctate; parts of the mouth, and base and apex of antennæ, reddish brown, the latter as long as the head and thorax, gradually thickened to the apex. Thorax transverse, narrowed in front, rounded at the sides, hinder margin nearly straight, posterior angles obtuse, thickly punctate with a depressed yellowish-grey pubescence. Elytra ovate, thickly punctate, obsoletely striate, clothed with a dense, fine, short, silvery-grey pubescence. Legs reddish brown. Length, 1¾-2 lines.
In decaying fungi, and carcases. Rare.

Fig. 100. NECRODES LITTORALIS, *L., Curtis.* [*F.* Silphidæ. *G.* Necrodes, *Leach.*]

Oblong, rather depressed, black. Head thickly and very finely punctate, club of antennæ ferruginous. Thorax orbiculate, with a faint central longitudinal channel, the sides depressed posteriorly, sparsely and finely punctate on the disc, rather thickly and coarsely at the sides and behind. Scutellum obtusely triangular, punctate. Elytra truncate at the apex, rather coarsely, thickly, and frequently confluently punctate; each with the suture and three longitudinal lines elevated, smooth. The males often have the posterior thighs incrassate, and the tibiæ bent. Length, 6-10 lines.
Common on carcases, especially on the coast.

Fig. 101. SILPHA OPACA, *L., Curtis.* [*F.* Silphidæ. *G.* Silpha, *L.*]

Ovate, depressed, black, opaque, clothed with a dense, yellowish, decumbent pubescence. Head thickly punctate. Thorax nearly semicircular, truncate in front, tri-sinuate behind, thickly punctate, with an obsolete longitudinal channel in the centre, and several depressions on each side. Elytra thickly punctate, the suture raised, the lateral margins reflexed, each with three elevated longitudinal lines. Length, 4½-5 lines.
In carcases. Not uncommon in the north; less frequent in the south.

Fig. 102. DIAPERIS BOLETI, *L., Curtis.* [*F.* Diaperidæ. *G.* Diaperis, *Geoffr.*]

Ovate, convex, black, shining. Head thickly and coarsely punctate, with a large impression in front. Thorax transverse, narrowed in front, rounded and margined at the sides, base bi-sinuate; finely and rather sparsely punctate. Elytra convex, with a broad, dentate, basal band, a narrow central one, and the apex externally bright orange ; each with eight rows of punctures, interstices exceedingly finely and sparsely punctate. Legs pitchy black, tarsi reddish brown. Length, 3-4 lines.
Very rare. In boleti at Barham, Suffolk ; near Hastings, Sussex ; and in Dalston Hall Wood, Cumberland.

Fig. 103. NECROPHORUS GERMANICUS, *L., Curtis.* [*F.* Silphidæ. *G.* Necrophorus, *Fab.*]

Oblong, black, reflexed margins of elytra red. Head finely and sparsely punctate, with an obtuse, triangular, yellow, membranous space in front ; antennæ, including the club, black. Thorax semicircular, anterior angles obtuse, lateral and posterior margins flat, coarsely punctate ; disc convex, finely and sparsely punctate, with a faint, central, longitudinal, impressed line, abbreviated at both ends, and a curved impression on each side. Scutellum thickly punctate. Elytra sinuate truncate at the apex, coarsely and sparsely punctate, the interstices exceedingly minutely punctate and reticulate. Length, 8-12 lines.
Exceedingly rare, but apparently widely distributed in the south.

23

PLATE XII.—*Continued.*

Fig. 104. SCAPHIDIUM QUADRIMACULATUM, *Oliv.*, *Curtis.* [*F.* Scaphidiidæ. *G.* Scaphidium, *Oliv.*]

Ovate, black, shining. Head sparsely and obsoletely punctate ; antennæ pitchy black, basal joints pitchy red. Thorax convex, transverse, narrowed in front, sides rounded anteriorly, nearly straight behind, base bi-sinuate, coarsely and rather sparsely punctate, base depressed, with a transverse bi-sinuate row of large oval punctures. Elytra ovate, truncate at the apex, coarsely and sparsely punctate ; each with a deep sutural stria curved at the base, and continued as far as the humeral callus, the sutural portion with remote, round punctures, the basal portion with large, oblong punctures, a curved, red spot at the base below the shoulders, and a transverse one near the apex. Legs black, tarsi pitchy red. Length, 2½–3 lines.
Not uncommon in rotten stumps of trees.

Fig. 105. STAPHYLINUS PUBESCENS, *De Geer, Curtis.* [*F.* Staphylinidæ. *G.* Staphylinus, *L.*]

Elongate, black, clothed with a dense, variegated pubescence. Head as wide or wider than the thorax, with a few large, remote punctures, clothed with a dense, long, yellowish-grey pubescence ; labrum testaceous ; mandibles, palpi, and antennæ black, basal joints of the latter reddish yellow at the base. Thorax elongate, narrowed behind, sides slightly rounded in front, base rounded, tessellated with brown and grey pubescence, with an interrupted, central, longitudinal ridge. Scutellum with two elongate patches of dense, black, velvety pubescence. Elytra and abdomen tessellated similarly to the thorax. Beneath pitchy black, abdomen with a dense silvery grey pubescence. Legs pitchy black, thighs with a testaceous ring near the apex. Length, 5–6½ lines.
Not uncommon, especially in the north.

Fig. 106. ANISOTOMA CINNAMOMEA, *Panzer.* (Male.) [*F.* Anisotomidæ. *G.* Anisotoma, *Illig.*] (Leiodes cinnamomea, *Curtis.*)

Oblong ovate, convex, reddish yellow or brown. Head minutely and thickly punctate, with a transverse row of four large shallow punctures on the forehead ; tips of mandibles and club of antennæ black. Thorax transverse, narrowed and emarginate in front, rounded at the sides, posterior margin nearly straight, rather thickly and finely punctate. Scutellum impressed in the centre, punctate. Elytra oblong, each with eight shallow, coarsely and closely punctate striæ ; interstices flat, exceedingly finely and sparsely punctate, the 1st, 3rd, 5th, and 7th with a few large, remote punctures. In the male the 2nd, 3rd, and 4th joints of the anterior tarsi are dilated, the intermediate thighs are toothed within at the base, and their tibiæ are curved, the posterior thighs are incrassate, with two large teeth at the apex, and their tibiæ curved ; in the female the anterior tarsi are simple, the intermediate and posterior thighs slender and unarmed, and their tibiæ only slightly bent. Length, 2–3½ lines.
Found in truffles, and by brushing towards sunset. It has occurred at Audley End, Essex ; Kimpton, Devon ; Mickleham, and near Croydon, Surrey ; in Suffolk, and Cornwall.

Fig. 107. EMUS HIRTUS, *L.*, *Curtis.* [*F.* Staphylinidæ. *G.* Emus, *Leach.*]

Elongate, black, with a dense, golden, silvery, and black pubescence. Head wide, covered with a dense, long, golden-yellow pubescence. Thorax nearly semicircular, clothed, except at its posterior margin, with a pubescence similar to that on the head. Scutellum with a dense, velvety, black pile. Basal third of elytra with a black pubescence, apical two-thirds occupied by a broad dentate band of silvery grey pubescence, intermixed with long black hairs. Abdomen with the three basal segments clad with short black hairs, the following three with a dense golden-yellow pubescence. Beneath with a violet tint, the sides of the 4th, and base of the 5th segment with a golden-yellow pubescence. Trochanters of the posterior legs of the male produced into a long curved spine, their tibiæ bent. Length, 7–10 lines.
Very rare. In dung, and flying in the hot sunshine. New Forest, and Parley Heath, Hants ; Whittlesea, Hunts ; Guildford, Surrey ; near Sheerness ; and in Norfolk and Devon.

PLATE XIII.

Fig. 108. OCYPUS (TASGIUS) PEDATOR *Grav.* [*F.* Staphylinidæ. *G.* Ocypus, *Leach.* *s-G.* Tasgius, *Steph.*] (Tasgius rufipes, *Curtis.*)

Elongate, sub-depressed, black ; head, thorax, and scutellum shining ; elytra and abdomen opaque ; antennæ, palpi, and legs red. Head a trifle narrower than the thorax, transverse, posterior angles rounded, thickly and coarsely punctate, with a smooth, central space. Thorax longer than broad, faintly narrowed posteriorly, sides slightly rounded, thickly and rather coarsely punctate, with a faint, smooth, longitudinal ridge in the centre. Scutellum thickly punctate. Elytra wider and a little longer than the thorax, very thickly and finely punctate, with a faint blue tint. Abdomen parallel, thickly and finely punctate, with a few large scattered punctures, and a black pubescence. Length, 6–9 lines.

Beneath stones. Tottenham ; Mickleham ; Folkestone ; Devonshire, &c.

Fig. 109. QUEDIUS LATERALIS, *Grav.*, *Curtis.* [*F.* Quediidæ. *G.* Quedius, *Steph.*]

Elongate, sub-depressed, black. Head sub-orbiculate, much narrower than the thorax, shining black, smooth, with a large puncture on the inner margin of each eye in front, and three placed in a triangle behind each eye ; mandibles. palpi, and base of antennæ pitchy red. Thorax transverse, narrowed in front, sides and base rounded in one unbroken curve ; convex, smooth, shining, with two dorsal series of three shallow punctures in front, and a few scattered punctures at the sides. Scutellum smooth, shining. Elytra a trifle longer than the thorax, sub-opaque, rather thickly and finely punctate, clothed with a sparse black pubescence, the reflexed margin pale yellow. Abdomen narrowed from the base to the apex, as finely but rather less thickly punctate than the elytra, iridescent. Legs pitchy black ; tarsi pitchy red, the anterior greatly dilated in the male. Length, 6–7 lines.

Rare, but widely distributed. Inhabits decayed fungi in fir woods.

Fig. 110. PHILONTHUS MARGINATUS, *Fab.*, *Curtis.* [*F.* Staphylinidæ. *G.* Philonthus, *Steph.*]

Elongate, black, shining, sides of thorax and legs reddish yellow. Head small, ovate, smooth, with two large approximate punctures placed transversely, and rather obliquely, near the inner margin of each eye in front, and eight or nine on each side behind ; palpi and antennæ with the basal joint testaceous, that of the latter with a black line above. Thorax scarcely as broad as long, narrowed in front, the sides and base continuously rounded, with two dorsal rows of four large punctures, and four others on each side. Scutellum and elytra black with a brassy green tint, finely and thickly punctate, clothed with a golden-yellow pubescence, the latter a little longer than the thorax. Abdomen with the sides parallel, black, sometimes faintly iridescent, sparsely and rather coarsely punctate, with a thin greyish pubescence ; beneath, with the apex of each segment narrowly red. Length, 4 lines.

Common in dung.

Fig. 111. CAFIUS FUCICOLA, *Curtis.* [*F.* Staphylinidæ. *G.* Cafius, *Curtis.*]

Elongate, pitchy black, sub-opaque, parts of the mouth and legs testaceous red. Head wider than the thorax, oblong quadrate, triangularly impressed in front, the sides and base with large irregularly disposed punctures ; apex of mandibles and palpi pitchy. Thorax nearly as wide as long, narrower than the elytra, narrowed behind, the sides slightly rounded in front, with two dorsal series of four large punctures, and four smaller ones on each side near the anterior angles. Scutellum thickly punctate. Elytra longer than the thorax, thickly and finely punctate, with a short grey pubescence. Abdomen with the sides parallel, coarsely and rather thickly punctate, sparsely pubescent ; each segment with two faint impressions in the centre near the base, and the anterior margin ferruginous. Anterior tarsi strongly dilated in the males. Length, 4–5 lines.

Rare. Beneath rejectamenta on the coast.

Fig. 112. ACHENIUM HUMILE, *Nicolai.* [*F.* Pæderidæ. *G.* Achenium, *Curtis.*] (A depressum, *Curtis*, but not of *Gravenhorst.*)

Elongate, flat, pitchy black, or brown, shining. Head as wide behind as the thorax, with a tubercle on each side at the base of the antennæ, sparsely and minutely punctate ; antennæ longer than the head and thorax, red. Thorax longer than broad, gradually narrowed to the base, pos-

E   25

PLATE XIII.—*Continued.*

terior angles rounded, reddish brown, sparsely and finely punctate, with a broad longitudinal, nearly smooth central line, and an oblong, slightly elevated, smooth space on each side of it towards the base. Elytra scarcely longer than the thorax, testaceous red, with a triangular black patch at the base round the scutellum, and a dusky spot at the outer apical angles, sparsely punctate, the punctures here and there disposed in indistinct rows. Abdomen parallel at the sides, finely and rather thickly punctate, apex, and frequently the apical margin of each segment ferruginous, with a thin short yellowish-grey pubescence. Legs testaceous red. Length, 3–3½ lines.

Rare ; Tottenham ; Highgate ; near Croydon and Reigate, Surrey, &c. In damp places.

Fig. 113. LATHROBIUM TERMINATUM, *Grav.*, *Curtis.* [*F.* Piederidæ. *G.* Lathrobium, *Grav.*]

Elongate, black, shining, abdomen sub-opaque. Head small, ovate, rather thickly and finely punctate : antennæ as long as the head and thorax, slender, pitchy black, red at the base and apex. Thorax narrower than the elytra, longer than broad, sub-quadrate, sides slightly rounded, all its angles obtuse, coarsely and thickly punctate, with a central longitudinal smooth space. Elytra longer than the thorax, thickly and finely punctate, with a bright red spot, in rare instances wanting, at the apex. Abdomen slightly dilated towards the middle, thence narrowed to the apex, very thickly and finely punctate. Legs testaceous red. Beneath ; the male has the 7th abdominal segment deeply notched at the apex, in the female the apex is acutely produced. Length, 3 lines.

Not uncommon. At the roots of grass and rushes in marshy places.

Fig. 114. CONOSOMA LITTOREUM, *L.* [*F.* Tachyporidæ. *G.* Conosoma, *Kraatz.*] (Tachyporus littoreus, *Curtis.*)

Oblong, broad in front, acuminate behind, pitchy black, clothed with fine short silky pubescence. Head triangular, convex, black, finely and sparsely punctate ; parts of the mouth and base of antennæ testaceous yellow, apex of the latter ferruginous. Thorax transverse, narrowed in front, sides rounded : posterior angles acute, salient ; thickly and very finely punctate, reddish brown, with a large ill-defined reddish-yellow patch at the hinder angles. Elytra longer than the thorax, similarly punctate, each with a large oblique reddish-yellow spot at the base. Abdomen narrowed from the base to the apex, immarginate, apical margin of each segment narrowly reddish brown. Legs reddish yellow. Length, 1¾–2 lines.

Common. In refuse of faggot-stacks, and beneath dead leaves.

Fig. 115. DELEASTER DICHROUS, *Grav.* [*F.* Oxytelidæ. *G.* Deleaster, *Eric.*] (Lesteva Leachii, *Curtis.*)

Rather elongate, sub-depressed, testaceous red, head black, abdomen pitchy black. Head triangular, a trifle wider than the thorax, with a tubercle on each side at the base of the antennæ, and a transverse impressed line between them, convex in front, with a transverse furrow behind, and two deep longitudinal impressions running obliquely into it : parts of the mouth and antennæ testaceous red, the latter dusky towards the apex. Thorax scarcely half as wide as the elytra, short, transverse, narrowed behind, sides rounded in front, base truncate, convex, finely and sparsely punctate, with an indistinct dorsal channel, and three large foveæ, one in the centre at the base, and one on each side. Elytra wide, flat, as long as the thorax, thickly and finely punctate, with a short golden-yellow pubescence. Abdomen with the sides parallel and broadly margined, the margin reflexed, very finely and sparsely punctate. Legs testaceous red. Length, 2¾–3¼ lines.

Rare. Under stones in wet places ; near Edinburgh ; Glasgow ; Croydon ; Colney Hatch, &c.

Fig. 116. SYNTOMIUM ÆNEUM, *Müll.* [*F.* Oxytelidæ. *G.* Syntomium, *Curtis.*] (S. nigroæneum, *Curtis.*)

Oblong, convex, black with a brassy hue. Head thickly and coarsely punctate, with a transverse impression between the antennæ, terminating on each side in a small fovea ; antennæ nearly as long as the head and thorax, black, their triarticulate club pitchy. Thorax convex, narrower than the elytra, transverse, narrowed posteriorly : sides rounded in front, sub-sinuate behind, finely serrated ; base sinuate, posterior angles acute, prominent : coarsely and sparsely punctate, with a central longitudinal smooth space, and a shallow depression on each side at the base. Elytra convex, longer than the thorax, very coarsely and ruggedly punctate. Abdomen short, the lateral margins wide and reflexed, shining, with a few indistinct punctures towards the sides. Legs pitchy red. Length, ⅞ line.

Not uncommon. In moss on shady banks.

26

PLATE XIV.

Fig. 117. PHYTOSUS NIGRIVENTRIS, *Chevr.* [*F.* Aleocharidæ. *G.* Phytosus, *Curtis.*]
(P. spinifer ♂, *Curtis.*)

Elongate, reddish yellow, clothed with a dense short grey pubescence, sub-opaque. Head globose, as wide as the thorax, dark brown, ferruginous at the base; parts of the mouth and antennæ pale yellow, the latter ferruginous at the apex. Thorax as broad as long, narrowed behind, rounded at the sides in front, with a distinct central longitudinal channel, finely punctate. Elytra conspicuously shorter than the thorax, shoulders rounded, apex emarginate, thickly and finely punctate. Abdomen dilated towards the apex, the 4th, 5th, and base of the 6th segments black. Legs testaceous, anterior and intermediate tibiæ thickly spinulose externally. Length, 1¼ line.
Rare ; on the coast. Beneath stones below high-water mark.

Fig. 118. PHYTOSUS SPINIFER, *Curtis.* [*F.* Aleocharidæ. *G.* Phytosus, *Curtis.*]
(P. spinifer, ♀, *Curtis.*)

Elongate, dull black, thickly and minutely punctate, clothed with a dense short grey pubescence, sub-opaque. Head narrower than the thorax, globose ; antennæ reddish yellow, parts of the mouth ferruginous, palpi dusky. Thorax narrower than the elytra, slightly transverse, narrowed behind, rounded at the sides in front, with a broad, shallow, central, longitudinal furrow. Elytra conspicuously longer than the thorax, shoulders prominent, apex emarginate, with an ill-defined ferruginous spot at the sutural angle. Abdomen very slightly dilated towards the tip, apex of the 6th segment and the 7th entirely reddish brown. Legs reddish yellow, anterior and intermediate tibiæ spinulose externally. Length, 1¼ line.
Very rare ; on the coast, in company with the preceding, near Ryde, Isle of Wight, and near Berwick-upon-Tweed.

Fig. 119. PROGNATHA QUADRICORNIS, *Kirby.* (Male.) [*F.* Piestidæ. *G.* Prognatha, *Latr.*] (Siagonium quadricorne, *Curtis.*)

Elongate, depressed, pitchy black, shining. Head as wide as the thorax, sparsely and finely punctate, forehead concave, produced on each side in front at the base of the antennæ into an erect flat horn, each mandible emitting from its base an elongate falcate process ; antennæ two-thirds of the length of the body, reddish brown. Thorax transverse, a trifle narrower than the elytra, narrowed behind, the sides gently rounded anteriorly, base truncate, posterior angles obtuse, finely and rather sparsely punctate. Elytra half as long again as the thorax, castaneous, each with a deep sutural, and four shallow, abbreviated, curved, dorsal striæ, the striæ finely and remotely punctate. Abdomen with the sides parallel, thickly and finely punctate, apex ferruginous. Legs reddish brown. Length, 2–2½ lines.
The female has the head narrower than the thorax, simply tuberculated on each side at the base of the antennæ, the mandibles unarmed, and the antennæ shorter.
Common in the south. Beneath bark of felled trees, especially elms.

Fig. 120. BLEDIUS TAURUS, *Germar.*, *var.* (Male.) [*F.* Oxytelidæ. *G.* Bledius, *Leach.*] (B. Skrimshirii, *Curtis.*)

Elongate, sub-cylindrical, black, thinly pubescent, elytra pale yellow, with a triangular black patch at the base. Head narrower than the thorax, faintly coriaceous, the forehead deeply impressed, armed with a long, erect, flat horn on each side at the base of the antennæ ; parts of the mouth pitchy red. Thorax transverse, as wide in front as the elytra, narrowed behind, sides straight to about the middle, thence abruptly rounded, base truncate, posterior angles very obtuse ; finely coriaceous, sparsely and coarsely punctate, the punctures shallow, with a fine but very distinct central longitudinal impressed line ; the anterior margin produced in the centre into an elongate horizontal spine, lying between the cephalic horns, and adorned on each side near the apex with a tuft of golden-yellow hairs. Elytra conspicuously longer than the thorax, rather thickly and coarsely punctate, the punctures very shallow, either entirely black, or pale yellow with a triangular black patch at the base. Abdomen with the sides parallel, finely coriaceous, the four basal segments sparsely punctate. Length, 2½–3½ lines.
The female has the forehead less deeply impressed, the lateral horns reduced to acute ridges. and the thorax straightly truncate in front, unarmed.
Rare ; near Fakenham, on the coast of Norfolk.

27

PLATE XIV.—*Continued.*

Fig. 121. **CALLICERUS OBSCURUS**, *Grav.* (Male.) [*F.* Aleocharidæ. *G.* Callicerus, *Grav.*] (♂ Callicerus Spencii, *Curtis.* ♀ C. hybridus, *Curtis.*)

Elongate, sub-depressed, pitchy black, with a sparse yellowish-grey pubescence, opaque. Head transverse, sub-orbiculate, narrower than the thorax, exceedingly obscurely punctate, with a deep impression in front between the antennæ, and a longitudinal impressed line behind it ; antennæ reddish brown, longer than the head and thorax, robust, of the male the 10th and 11th joints elongate, of the female the 10th scarcely longer than the 9th. Thorax as broad as long, narrowed in front, sides rounded anteriorly, base truncate, very obscurely punctate, with a faint central longitudinal impression behind. Elytra conspicuously longer than the thorax, brown, thickly and faintly punctate, sub-opaque. Abdomen with the sides parallel, shining, smooth above, obscurely punctate beneath. Legs testaceous red, thighs sometimes pitchy. Length, 1¼-1⅓ lines.

Rare. In marshy places ; banks of the Thames near Hammersmith ; near Croydon, Surrey; &c.

Fig. 122. **OXYPORUS MAXILLOSUS**, *Fab., Curtis.* [*F.* Oxytelidæ. *G.* Oxyporus, *F.*]

Oblong, rather convex, smooth, shining, reddish yellow, head, thorax, and outer apical angles of elytra black. Head transverse quadrate, as wide or wider than the thorax, very finely and sparsely punctate, with a large triangular impression in front between the antennæ. Thorax half as wide as the elytra, transverse, narrowed behind, the sides slightly rounded, smooth, impunctate, with a broad transverse impression a little before the middle. Elytra half as long again as the thorax, each with a deep sutural punctate stria, and two broad abbreviated, irregularly punctate, longitudinal impressions on the disc. Abdomen narrowed from the base to the apex, with a broad reflexed margin, impunctate. Length, 3-4 lines.

Inhabits fungi ; " near Cheltenham and Bristol," and " in Suffolk and Devonshire."

Fig. 123. **HYGRONOMA DIMIDIATA**, *Grav.* [*F.* Aleocharidæ. *G.* Hygronoma, *Eric.*] (Homalota dimidiata, *Curtis.*)

Elongate, depressed, densely punctate, opaque, black, palpi, base of antennæ, apical two-thirds of elytra, and legs reddish yellow. Head transverse, nearly as wide as the thorax, constricted behind, depressed, with a broad central longitudinal furrow behind. Thorax a little broader than long, slightly narrowed behind, sides rounded, with a broad central longitudinal impression. Elytra a little wider and conspicuously longer than the thorax. Abdomen with the sides parallel, the basal segments transversely impressed behind. Length, 1¼ lines.

Inhabits marshy places amongst reeds, in the stems of which it hybernates.

Fig. 124. **DIANOUS COERULESCENS**, *Gyll., Curtis.* [*F.* Stenidæ. *G.* Dianous, *Leach.*]

Elongate, sub-cylindrical, shining, black with a deep blue tint, elytra with a round red spot. Head transverse, sub-orbiculate, wider than the thorax, thickly and finely punctate, with two broad longitudinal furrows between the eyes ; antennæ and palpi black, the latter pitchy at the apex. Thorax elongate, sub-cylindrical, slightly narrowed behind ; sides rounded in front, sinuate behind : thickly and rather coarsely punctate. Elytra longer than the thorax, convex, thickly and rather coarsely punctate, each with a large circular orange-yellow spot surrounded by a violet-coloured ring, a little behind the middle. Abdomen narrowed from the base to the apex, very finely and thickly punctate. Legs long, pubescent. Length, 2¼-2¾ lines.

Widely distributed, but not common. Inhabits moss in running water.

Fig. 125. **DINARDA MAERKELII**, *v. Kiesenw.* [*F.* Aleocharidæ. *G.* Dinarda, *Leach.*] (Lomechusa dentata, *Curtis,* but not of *Fabricius.*)

Oblong, broad and obtuse in front, acuminate behind, depressed, pitchy black, pubescent, opaque, sides of thorax and elytra red. Head small round, forehead depressed, thickly and rather coarsely punctate : antennæ pitchy black, red at the base and apex. Thorax rather wider than the elytra, transverse, narrowed and emarginate in front, rounded at the sides, base deeply bi-sinuate, posterior angles acute, lateral margins flat, disc convex with a faint central longitudinal impressed line, rather coarsely and thickly punctate. Elytra as long as the thorax, similarly punctate, sinuate truncate at the apex, outer angles acute. Abdomen narrowed from the base to the apex, lateral margins broad, reflexed, coarsely and sparsely punctate. Legs red. Length, 2 lines.

Inhabits the nests of the wood-ant (Formica rufa).

PLATE XV.

Fig. 126. PÆDERUS FUSCIPES, *Curtis.* [*F.* Pæderidæ. *G.* Pæderus, *Grav.*]

Elongate, sub-cylindrical. Head black, nearly as wide as the thorax, sub-orbiculate, smooth in front, coarsely and sparsely punctate at the sides and posteriorly, with a transverse punctate impression between the eyes, and an elongate fovea on each side in front of them ; mandibles and maxillary palpi reddish yellow, the latter with the apex of the third joint fuscous, labial palpi dusky, antennæ black, three basal joints reddish yellow. Thorax oblong, convex, much narrower than the elytra, sides nearly straight, red, smooth, with a few irregularly disposed punctures and long black hairs at the sides. Scutellum black. Elytra half as long again as the thorax, blue, with a sparse silvery-grey pubescence, thickly and coarsely punctate. Abdomen with the sides parallel, narrowly margined, finely and sparsely punctate, red, the two apical segments black. Breast black. Legs reddish yellow, apex of thighs black, tibiæ and tarsi fuscous. Length, 3 lines.

Rare. New Forest ; near Ventnor, Isle of Wight ; &c.

Fig. 127. STENUS GUTTULA, *Müll.* [*F.* Stenidæ. *G.* Stenus, *Latr.*] S. Kirbii, *Curtis.*)

Elongate, sub-cylindrical, black, with a dense short silvery-grey pubescence. Head transverse, much wider than the thorax, thickly and coarsely punctate, deeply impressed in front, with a broad, deep, longitudinal furrow on each side between the eyes, the intermediate space convex, smooth ; palpi testaceous, the third joint dusky at the apex ; antennæ pitchy, basal joint black. Thorax elongate, sub-cylindrical ; sides rounded anteriorly, slightly sinuate behind ; thickly and coarsely punctate, interstices uneven, the disc with two small, elongate, raised, smooth spaces. Elytra a trifle longer than the thorax, thickly, deeply, and coarsely punctate, each with a reddish-yellow circular spot a little behind the middle. Abdomen narrowed from the base to the apex, narrowly margined, rather thickly and finely punctate. Legs long, slender, pale yellow, apical half of femora black, base and apex of tibiæ pitchy black ; tarsi elongate. Length, 2–2¼ lines.

Rare. Tottenham, Essex ; Charlton, Kent ; Isle of Wight ; Bristol ; &c.

Fig. 128. STILICUS FRAGILIS, *Grav.* [*F.* Pæderidæ. *G.* Stilicus, *Latr.*] (Rugilus fragilis, *Curtis.*)

Elongate, black, thorax, scutellum, and anterior legs red. Head orbiculate, much wider than the thorax, convex, black, very thickly, finely, and ruggedly punctate, opaque ; palpi and antennæ pitchy black, apex of the latter ferruginous. Thorax oblong, attenuated in front, narrowed behind, sides rounded, convex, blood red, very thickly and finely punctate, with a central longitudinal slightly elevated, smooth space. Scutellum red. Elytra of the width of the head, a little longer than the thorax, black, with the apex narrowly testaceous, slightly shining, thickly and finely punctate, with a dense short yellowish-grey pubescence. Abdomen margined, gradually widened from the base to the fourth segment, thence narrowed to the apex, black, very finely and thickly punctate, with a similar pubescence to that on the elytra. Anterior legs red, intermediate pitchy red, posterior pitchy black. Length, 3 lines.

Rare, although very widely distributed. Found beneath dead leaves, in the refuse of hay and faggot stacks, and in moss in damp, shady places.

Fig. 129. FALAGRIA THORACICA, *Steph., Curtis.* [*F.* Aleocharidæ. *G.* Falagria, *Leach.*]

Elongate, reddish brown, shining. Head transverse, sub-orbiculate, very nearly as wide as the thorax, pitchy red, very faintly punctate, with a shallow depression on the forehead ; antennæ pitchy red, the basal joints ferruginous. Thorax as broad as long, narrowed behind ; sides rounded in front, slightly sinuate posteriorly ; red, indistinctly punctate, with a deep central longitudinal channel, terminating behind in a triangular fovea. Scutellum very finely and thickly punctate. Elytra oblong quadrate, a little longer than the thorax, reddish brown, very finely and thickly punctate, and clothed with a thin grey pubescence. Abdomen slightly dilated posteriorly, reddish brown, as finely but less thickly punctate than the elytra. Legs ferruginous. Length, 1¼ lines.

Not common, although very widely distributed. Found in moss, at roots of grass, and beneath dead leaves, often in the company of ants, especially of Formica flava and Myrmica rubra.

29

PLATE XV.— Continued.

Fig. 130. TRICHONYX MÆRKELII, *Aubé.*—[*F.* Pselaphidæ. *G.* Trichonyx, *de*
Chaudoir.] (Bryaxis sulcicollis, *Curtis*, but not of *Reichenbach.*)

Oblong, reddish yellow or castaneous, shining, sparsely clothed with a yellowish-grey pubescence. Head triangular, a little narrower than the thorax, with a deep horse-shoe shaped impression in front, crown convex, with a small, round, shallow fovea behind. Thorax as wide as long,
narrowed in front and behind, sides rounded anteriorly; with a deep central longitudinal channel,
terminating behind in a triangular impression, connected by a faint arcuate impressed line with a
large fovea situate on each side a little behind the middle. Elytra half as long again as the thorax,
shoulders rounded, dilated behind, sinuate truncate at the apex, the outer angles acuminate,
obsoletely punctate, with a fine deep sutural stria, and a broad abbreviated one on the disc.
Abdomen smooth. Legs reddish yellow. Length, 1 line.
Very rare. Carlisle, "taken off the city walls;" Mickleham, Surrey, in nests of Formica flava,
beneath flints.

Fig. 131. BYTHINUS PUNCTICOLLIS, *Denny.* [*F.* Pselaphidæ. *G.* Bythinus, *Leach.*]
(Arcopagus puncticollis, *Curtis.*)

Oblong, reddish brown, shining, with a sparse yellowish-grey pubescence. Head narrower
than the thorax, triangular, coarsely and sparsely punctate; palpi and antennæ reddish yellow,
the basal joint of the latter cylindrical in the female, swollen in the middle and angular within at
the apex in the male. Thorax transverse, convex, narrowed in front and behind, the sides rounded
anteriorly; coarsely and sparsely punctate, with a faint transverse impressed line at the base.
Elytra conspicuously longer than the thorax, dilated towards the apex, sinuate truncate at the tip,
pitchy black, very coarsely and sparsely punctate, each with a sutural stria, and an oblong impression within the shoulder. Abdomen very finely punctate. Legs reddish yellow; thighs thickened,
posterior tibiæ curved in the male. Length, ⅔ line.
Not uncommon. In moss and at roots of grass in humid places.

Fig. 132. HOLOPARAMECUS SINGULARIS, *Beck.* [*F.* Lathridiidæ. *G.* Holoparamecus, *Curtis.*] (H. depressus, *Curtis.*)

Oblong, depressed, reddish yellow, shining. Head triangular. Thorax elongate, narrowed
behind, sides rounded anteriorly, exceedingly finely punctate, with a deep elongate fovea on each
side at the base, connected by a transverse impressed line. Elytra oblong, very finely punctate and
pubescent, with a fine sutural stria obliterated towards the apex. Length, ⅓ line.
Occasionally found in shops and warehouses, probably introduced in foreign merchandise.

Fig. 133. LATHRIDIUS ELONGATUS, *Curtis.* [*F.* Lathridiidæ. *G.* Lathridius, *Hbst.*]

Elongate, red brown. Head oblong, quadrate, with a shallow longitudinal impression on each
side between the antennæ, thickly and finely rugulose-punctate; antennæ reddish yellow. Thorax
longer than broad, slightly narrowed behind; sides faintly rounded anteriorly, sinuate behind;
margined; posterior angles obtuse, thickly and finely rugulose-punctate, with a transverse impression at about one-third from the base. Elytra elongate ovate, the suture raised, each with six
broad punctate striæ, the punctures very large and approximate; interstices narrow, smooth.
Length, ⅔ line.
Rare. In moss on trunks of trees, and in boleti; Colney Hatch and Finchley, near London;
New Forest; Bristol; &c.

Fig. 134. PARAMECOSOMA MELANOCEPHALUM, *Hbst.* [*F.* Cryptophagidæ.
G. Paramecosoma, *Curtis.*] (P. bicolor, *Curtis.*)

Elongate, ovate, pitchy black, antennæ, elytra and legs reddish brown, sparsely clothed with a
fine short grey pubescence. Head narrower than the thorax, convex, rather thickly punctate.
Thorax transverse, narrowed in front and behind; sides slightly rounded, with two obscure denticulations equidistant from each other and from the anterior and posterior angles; rather thickly
and finely punctate. Elytra oblong ovate, thickly punctate, the punctures arranged in irregular
rows at the base, obsolete at the apex, sutural stria fine, deeply impressed at the apex, obscure at
the base. Length, ⅔ line.
Not uncommon in the north, apparently rare in the south. Colney Hatch and Finchley, near
London; "in some abundance in June, off furze-bushes growing over the water near Knaresborough,
in Yorkshire."

36

PLATE XVI.

Fig. 135. DITOMA CRENATA, *Hbst.* [*F.* Synchitidæ. *G.* Ditoma, *Illiger.*] (Bitoma crenata, *Curtis.*)

Oblong, parallel, depressed, black, opaque, elytra each with two large red spots. Head transverse, narrower than the thorax, rugulosely punctate, nearly smooth in front, with a large, deep, oblique impression on each side between the antennæ; antennæ red brown, apical joint paler. Thorax transverse, sub-quadrate, slightly narrowed behind, faintly rounded at the sides anteriorly, base sinuate, posterior angles obtuse; rather thickly rugulose-punctate; sides narrowly margined, crenulate; disc depressed, with two longitudinal ridges on each side, the inner one curved, the outer one nearly straight. Elytra with the sides nearly parallel, punctate-striate, the punctures quadrate, suture and alternate interstices raised, each with a large red patch at the base and another at the apex. Legs red, thighs black. Specimens frequently occur of a uniform ferruginous colour throughout, and are probably immature. Length, 1½–1¾ lines.

Not uncommon. Beneath bark.

Fig. 136. RHIZOPHAGUS BIPUSTULATUS, *Fab., Curtis.* [*F.* Nitidulidæ. *G.* Rhizophagus, *Herbst.*]

Elongate, sub-depressed, black, shining, legs, antennæ, and a more or less conspicuous spot near the apex of each elytron red. Head a little narrower than the thorax, rather thickly and finely punctate, with a shallow oblique impression on each side between the antennæ. Thorax scarcely longer than broad, slightly narrowed behind, the sides nearly straight, somewhat sparsely punctate, the punctures oblong. Elytra punctate-striate; interstices flat, smooth, that next the suture with a row of exceedingly minute punctures. Length, 1¼–1½ lines.

Common. Under bark, and in boleti.

Fig. 137. HYPOPHLŒUS BICOLOR, *Olivier, Curtis.* [*F.* Ulomidæ. *G.* Hypophlœus, *Fab.*]

Elongate, sub-cylindrical, shining, ferruginous, eyes and posterior two-thirds of elytra black. Head small, narrower than the thorax, very finely and rather thickly punctate, with a transverse arcuate impressed line in front of the eyes. Thorax oblong, convex, narrowed in front and behind, the sides moderately rounded and narrowly margined, posterior angles obtuse, finely and rather thickly punctate. Elytra with the sides nearly parallel, nearly as finely and not quite so thickly punctate as the thorax. Length, 1¾–2 lines.

Common in the south. Beneath the bark of felled elms.

Fig. 138. TENEBRIO OBSCURUS, *Fab., Curtis.* [*F.* Tenebrionidæ. *G.* Tenebrio, *L.*]

Elongate, sub-depressed, black, opaque. Head narrower than the thorax, coarsely, thickly, and confluently punctate, with a faint, transverse, arcuate, impressed line in front of the eyes. Thorax as broad as long, slightly narrowed in front, sides rounded; base sinuate, with a short longitudinal linear impression on each side; coarsely, thickly, and confluently punctate. Elytra with the sides nearly parallel, striate, the striæ obscurely and remotely punctate; interstices rather convex, thickly rugulosely punctate. Legs pitchy. Length, 6–8 lines.

Common. In bake-houses and stables near London.

PLATE XVI.—*Continued.*

Fig. 139. SARROTRIUM CLAVICORNE, *L.* [*F.* Colydiidæ. *G.* Sarrotrium, *Illiger.*]
(S. muticum, *Curtis.*)

Oblong, black, opaque, with a fine ashy pubescence. Head narrower than the thorax, sub-quadrate, impressed in front, rugulose-punctate; antennæ with long erect black hairs. Thorax transverse, slightly narrowed in front, sides straight, disc with two elevated longitudinal ridges, and a deep, broad, posteriorly sub-interrupted, central channel; rugulose-punctate. Elytra coarsely punctate-striate, the 2nd, 4th, and 6th interstices raised. Length, 1¼-2 lines.
Frequents sandy places.

Fig. 110. ALPHITOBIUS PICEUS, *Olivier.* [*F.* Ulomidæ. *G.* Alphitobius, *Steph.*]
(Uloma fagi, *Curtis.*)

Oblong, pitchy black, shining, anterior portion of the head, parts of the mouth, antennæ, and legs reddish brown. Head small, transverse, finely and thickly punctate, with a transverse arcuate impression between the antennæ. Thorax transverse, narrowed in front, and slightly so behind, sides rounded, narrowly margined, the margin reflexed, base bi-sinuate, very narrowly margined throughout; posterior angles nearly rectangular; rather finely and thickly punctate throughout. Scutellum almost semicircular, finely and thickly punctate. Elytra with the sides nearly parallel, punctate-striate; interstices slightly convex, finely and thickly punctate. Legs reddish brown; anterior tibiæ nearly straight, slightly and gradually dilated from the base to the apex, their external margin with a few obsolete denticulations. Length, 2½-3 lines.
Not uncommon in bake-houses.

Fig. 141. ELEDONA AGRICOLA, *Hbst.* [*F.* Bolitophagidæ. *G.* Eledona, *Latr.*]
(Bolitophagus agricola, *Curtis.*)

Oblong, convex, pitchy black, opaque. Head small, finely and very thickly punctate, nearly smooth in front, with an obsolete transverse impression between the antennæ; palpi and antennæ reddish brown. Thorax transverse, convex, slightly narrowed in front, sides nearly straight, crenulate; thickly and coarsely reticulate-punctate. Elytra with the sides parallel, convex, each with nine broad coarsely punctate longitudinal furrows; interstices narrow, smooth. Length, 1¼-1½ lines.
Not common. In boleti; Epping Forest; Richmond Park; Kensington Gardens; Netley, Shropshire; Norfolk; Suffolk; &c.

Fig. 142. HELOPS PALLIDUS, *Curtis.* [*F.* Helopidæ. *G.* Helops, *Fab.*]

Oblong ovate, convex, pale testaceous yellow. Head small, transverse, very finely and thickly punctate, with an arcuate impression between the antennæ. Thorax transverse, narrowed in front, very slightly so behind, sides rounded anteriorly, a little less finely and thickly punctate than the head. Scutellum broadly triangular, thickly punctate. Elytra ovate, convex, finely punctate-striate; interstices flat, exceedingly finely and sparsely punctate. Length, 3-5 lines.
At roots of grass in sandy situations on the coast; Barmouth, Merioneth; Tenby, Pembroke; Swansea; &c.

Fig. 143. MICROZOUM TIBIALE, *Fab.* [*F.* Opatridæ. *G.* Microzoum, *Redtb.*]
(Opatrum tibiale, *Curtis.*)

Oblong, convex, black, sub-opaque. Head transverse, coarsely and thickly punctate, with a shallow arcuate impression between the eyes. Thorax transverse; sides rounded in front, sinuate behind; base indistinctly bi-sinuate, posterior angles rectangular; coarsely and thickly punctate, the disc with three slightly raised smooth spaces. Elytra oblong, convex, rather coarsely and thickly punctate, with several indistinct longitudinal furrows and foveæ. Legs pitchy black, anterior tibiæ triangularly enlarged. Length, 1½-2 lines.
Not uncommon in sandy situations, especially on the coast.

32

PLATE XVII.

Fig. 144. LAGRIA HIRTA, *L.*, *Curtis*. (Male.) [*F*. Lagriidæ. *G*. Lagria, *Fab*.]

Elongate, black, rather sparsely clothed with fine long sub-erect yellow hairs, elytra brownish yellow. Head small, sub-orbiculate, rather thickly and coarsely punctate, eyes large and approximate, antennæ long and slender, the apical articulation longer than the three preceding joints united. Thorax elongate, narrowed in front, coarsely and sparsely punctate, with a transverse furrow about one-third from the base. Elytra slightly dilated at the sides, thickly and finely rugulose-punctate, with indistinct striæ. In the female the eyes are much smaller and widely separate, the head and thorax are more thickly punctate, the antennæ are shorter and thicker, and the apical joint is but a little longer than the two preceding united, and the elytra are more dilated at the sides. Length, 3–4½ lines.

Common on flowers, and amongst rank herbage, towards the end of summer.

Fig. 145. CISTELA CERAMBOIDES, *L.*, *Curtis*. [*F*. Cistelidæ. *G*. Cistela, *Fab*.]

Elongate, elliptical, black, with a fine sparse ferruginous pubescence, elytra reddish yellow. Head triangular, thickly punctate, forehead with a shallow fovea ; antennæ three-fourths of the length of the body, acutely serrated within in the males, a little shorter and obtusely serrated in the females. Thorax transverse, narrowed in front, bi-sinuate at the base ; very thickly and finely rugulose-punctate. Elytra finely punctate-striate, interstices rather convex, exceedingly minutely rugulose-punctate. Length, 4½–5¼ lines.

Not uncommon in the south. In flowers, and on decayed trees.

Fig. 146. OMOPHLUS ARMERIÆ, *Curtis*. [*F*. Cistelidæ. *G*. Omophlus, *Solier*.]

Elongate, convex, black, shining, elytra reddish yellow. Head triangular, punctate, with a fovea in front and a few long black hairs behind ; base of antennæ and parts of the mouth pitchy brown. Thorax transverse, narrowed behind, with a shallow, interrupted, central, longitudinal channel, and a deep fovea on each side ; finely punctate, sparsely clothed with decumbent grey and sub-erect black hairs. Scutellum thickly punctate. Elytra punctate-striate ; interstices convex, thickly and coarsely punctate. Length, 3½–4 lines.

Local ; near Weymouth, and on the Chesil Bank in the Isle of Portland. On the flowers of the common thrift (*Statice Armeria*).

Fig. 147. BLAPS LETHIFERA, *Marsham*. [*F*. Blapsidæ. *G*. Blaps, *Fab*. CHURCH-YARD BEETLES.] (B. obtusus, *Curtis*.)

Elongate ovate, black, slightly shining. Head rather thickly and finely punctate, with a transverse faintly impressed curved line between the antennæ ; antennæ short, not reaching to the base of the thorax, robust, the 5th and 6th joints very little longer than broad. Thorax transverse, narrowed in front, very slightly so behind ; the sides rounded anteriorly ; rather thickly and finely punctate. Elytra thrice the length of the thorax, their greatest width at one-fourth of their length from the base, terminated by a short caudal appendage, rather sparsely and coarsely rugulose-punctate, presenting faint traces of longitudinal ridges. In the male, the anterior margin of the basal segment of the abdomen has in the centre an oblong tubercle densely fringed with ferruginous hairs. Length, 10–12 lines.

Common in cellars and stables.

Fig. 148. MELANDRYA CANALICULATA, *Fab.*, *Curtis*. [*F*. Melandryidæ. *G*. Melandrya, *Fab*.]

Elongate, black ; palpi, apical joint of antennæ, and tarsi, reddish yellow. Head finely punctate, with an elongate impression in front. Thorax transverse, narrowed in front, rounded at the sides, base bi-sinuate, with a deep central longitudinal furrow, and a still deeper elongate fovea on each

F

PLATE XVII.—*Continued.*

side at the base ; finely and thickly punctate. Elytra wider than the thorax, thickly and finely punctate, with a short depressed silky grey pubescence ; each with four broad, flat, longitudinal furrows abbreviated at the base. Length, 6 lines.

Very rare. In dead trees ; New Forest, near Brockenhurst.

Fig. 149. ORCHESIA UNDULATA, *Kraatz.* [*F.* Melandryidæ. *G.* Orchesia, *Latr.*] (O. fasciata, *Curtis,* but not of *Paykull.*)

Elongate ovate, reddish brown, minutely and thickly rugulose-punctate, clothed with a silky pubescence ; antennæ pitchy black in the middle, reddish yellow at the base and apex. Thorax transverse, narrowed in front, sides rounded, base bi-sinuate with a triangular impression on each side, and a more or less apparent transverse bi-arcuate black band on the disc. Elytra reddish yellow, with two oblique interrupted bands at the base, a broad transverse fascia towards the tip, and the apex black. Length, 2–2½ lines.

Rare. New Forest, near Brockenhurst ; in boleti, and on flowers.

Fig. 150. ABDERA BIFASCIATA, *Marsham.* [*F.* Melandryidæ. *G.* Abdera, *Steph.*] (Hypulus biflexuosus, *Curtis.*)

Oblong, pitchy black, finely and rather thickly rugulose-punctate, clothed with a silky yellowish-grey pubescence. Head transverse, deflexed, palpi and two basal joints of antennæ testaceous. Thorax a little longer than broad, narrowed in front, sides slightly rounded, base bi-sinuate, posterior angles obtuse ; with an exceedingly obscure central longitudinal impression behind. Elytra with two flexuous testaceous or pale yellow bands, one a little before the middle, the other at about one-third from the tip. Legs reddish brown, thighs sometimes pitchy black. Length, 1¼–1¾ lines.

Rare. In dead boughs of oak, and upon old palings ; Highgate ; Wimbledon ; near Reigate, &c.

Fig. 151. CARIDA FLEXUOSA, *Payk.* [*F.* Melandryidæ. *G.* Carida, *Mulsant.*] (Hallomenus flexuosus, *Curtis.*)

Oblong, bright ferruginous, head, a transverse bi-lobed patch on the thorax, scutellum, two zigzag bands on the elytra, and the abdomen black, very minutely and rather thickly rugulose-punctate, with a fine silky pubescence. Head deflexed, triangular ; antennæ brown in the middle, reddish yellow at the base and apex. Thorax transverse, narrowed in front, sides rounded, base bi-sinuate, posterior angles obtuse, with a faint central longitudinal impressed line. Length, 1½–2 lines.

Beneath bark of willow, and in fungi on old trees. Very rare. Near Peterborough, and in the vicinity of Cambridge.

Fig. 152. RHIPIPHORUS PARADOXUS, *L., Curtis.* [*F.* Rhipiphoridæ. *G.* Rhipiphorus, *Fab.*]

Elongate, black, with a short pubescence, sides of the thorax posteriorly, abdomen and elytra reddish yellow. Thorax elongate, narrowed in front, sides slightly rounded anteriorly, sinuate posteriorly ; base bi-sinuate, the central lobe produced over the scutellum ; posterior angles acute, embracing the humeral angles of the elytra ; with a broad and very deep central longitudinal excavation ; smooth in the centre ; sides punctate, each puncture emitting a short erect fuscous hair. Elytra nearly as wide at the base as the thorax, with a short depressed pubescence, gaping at the suture, narrowed to the apex, acuminate at the tip ; each with a broad longitudinal furrow ; of the males yellow with the tips black, of the females entirely black. Length, 5–9 lines.

Widely distributed, but not common. In wasps' nests, and occasionally in flowers.

34

PLATE XVIII.

Fig. 153. MORDELLISTENA ABDOMINALIS, *Fab.* [*F.* Mordellidæ. *G.* Mordellistena, *Costa.*] (Mordella abdominalis, *Curtis.*)

Elongate, pitchy black, with a silky pubescence, thorax and abdomen orange yellow. Head triangular, nutant, convex : parts of the mouth, and base and apex of antennæ pale yellow, the latter reddish brown in the middle. Thorax transverse, narrowed in front, sides slightly rounded ; base bi-sinuate, central lobe obtuse ; posterior angles rounded ; with a dense long depressed golden-yellow pubescence. Elytra nearly as broad at the base as the thorax, sides nearly parallel, exceedingly finely and thickly punctate, with a close short depressed brown pubescence. Abdomen reddish yellow ; pygidium acuminate black, reddish yellow at the base. Anterior legs reddish yellow ; intermediate reddish brown ; posterior pitchy black. Length, 2½-3¼ lines.
Not common ; but widely distributed in the south.

Fig. 154. SITARIS MURALIS, *Forster.** [*F.* Meloidæ. *G.* Sitaris, *Latr.*] (S. humeralis, *Fab.*,† *Curtis.*)

Elongate, black, slightly shining, base of elytra yellow. Head triangular, thickly and coarsely punctate behind, rather sparsely in front, interstices shining ; antennæ of the male about four-fifths of the entire insect, of the female a little longer than the head and thorax. Thorax sub-quadrate ; sides rounded in front, sinuate behind ; posterior angles obtuse ; very thickly and coarsely punctate on the disc, rather sparsely at the sides, central longitudinal channel deep behind, obsolete in front. Scutellum punctate at the base, nearly smooth at the apex. Elytra broader than the thorax at the base, rather coarsely and rugulosely punctate, pitchy black, the base yellow. Legs pitchy black, tibiæ and tarsi frequently reddish brown. Length, 5-6¼ lines.
Not common, but widely distributed in the south. Found on old walls inhabited by Anthophora retusa, in whose cells the larva resides.

Fig. 155. MELOE BREVICOLLIS, *Panzer, Curtis.* [*F.* Meloidæ. *G.* Meloe, *L.*]

Elongate, black, rather shining, with a blue tint. Head transverse, coarsely, deeply and sparsely punctate ; antennæ short, robust, thickened towards the apex. Thorax transverse, narrowed in front, slightly so behind, sides rounded, base broadly emarginate ; coarsely, deeply, and sparsely punctate, with a fine central longitudinal impressed line. Elytra coarsely, ruggedly, and confluently punctate. Abdomen above black, sub-opaque, exceedingly finely punctate : each segment having in the centre a slightly shining, bluish, thickly punctate space ; beneath shining blue-black, rather finely sparsely rugulosely punctate. Legs short, robust, claws pitchy. Length, 4-9 lines.
Rare ; near Ripley, Surrey ; Windsor ; Christchurch ; Tavistock ; and near Plymouth.

Fig. 156. OSPHYA BIPUNCTATA, *Fab.* (Female.) [*F.* Melandryidæ. *G.* Osphya, *Illiger.*] (Nothus bimaculatus, *Curtis.*)

Elongate, reddish yellow, clothed with a dense short grey pubescence, head and thorax rufous, two spots on the crown of the head, antennæ except the three basal joints, two large patches on the thorax, tips of elytra, of thighs, and of tibiæ, black. Head deflexed, punctate ; palpi dusky at the tip ; eyes deeply notched, black. Thorax transverse, all its angles and the sides rounded, base truncate, finely punctate. Elytra wider than the thorax, sides nearly parallel, thickly and finely punctate. Posterior thighs slender. Length, 4-4½ lines.
Rare. On whitethorn blossoms, Monk's Wood, Hunts ; Windsor ; and Weston-on-the Green.

Fig. 157. CONOPALPUS TESTACEUS, *Oliv., Curtis.* [*F.* Melandryidæ. *G.* Conopalpus, *Gyll.*]

Elongate, reddish yellow, rather thickly and minutely punctate, clothed with a sparse shining depressed yellowish-grey pubescence. Head small convex, deflexed, rather thickly and finely

* Necydalis muralis, *Forster, Nova Species Insectorum Centuria* I. 48 (1771).
† Necydalis humeralis, *Fab. Syst. Ent.* 209, 4 (1775).]

35

PLATE XVIII.—*Continued.*

punctate, with a transverse depression between the antennæ; antennæ and tips of mandibles black, the three basal joints of the former reddish yellow. Thorax transverse, narrowed in front, and slightly so behind; sides rounded; base bi-sinuate, with two small approximate foveæ in the centre opposite the scutellum; rather thickly and finely punctate. Elytra with the sides nearly parallel, finely rugulose-punctate. Legs ferruginous. Length, 3–3½ lines.
Not common, but widely distributed in the south. Sherwood Forest; Darenth Wood; Reigate; Kensington Gardens; New Forest; &c.

Fig. 158. OSPHYA BIPUNCTATA, *Fab.* (Male.) [*F.* Melandryidæ. *G.* Osphya, *Illiger.*] (Nothus clavipes, *Curtis.*)

Elongate, black, clothed with a fine dense grey pubescence; month, three basal joints of antennæ, margins of thorax and of elytra, base of thighs and of tibiæ, reddish yellow. Head deflexed, finely punctate; palpi with the apex dusky. Thorax transverse, rounded in front and at the sides, base truncate, angles rounded. Elytra with the sides parallel, thickly and finely punctate. Posterior thighs incrassate. Length, 3–4½ lines.
Rare. On whitethorn blossoms; Monk's Wood, Hunts: and near Windsor.

Fig. 159. CANTHARIS VESICATORIA, *L.*, *Curtis.* SPANISH FLY. BLISTER BEETLE.
[*F.* Cantharidæ. *G.* Cantharis, *Geoffr.*]

Elongate, shining, bright green, golden green, or blue. Head transverse, deflexed, triangular, minutely and rather sparsely punctate, with a very thin short erect golden-yellow pubescence, and a deep central longitudinal furrow; antennæ black, pubescent, basal joint glabrous, with a blue or green tint. Thorax transverse, narrowed behind, sides straight, anterior angles obtuse, posterior angles rounded, base emarginate and reflexed; anterior portion depressed: disc flat, with a distinct central longitudinal impressed line. rather more closely punctate than the head. Elytra with the sides parallel, finely shagreened, each with two distinct longitudinal elevated lines. Legs black, with a green tint, covered with a short golden-yellow pubescence, claws amber yellow. Beneath bright blue or green, thickly punctate. breast with a long yellowish-grey pubescence. Length, 5–11 lines.
Rare, but occasionally found in abundance; near Norwich; Cheltenham; Isle of Wight; Cambridgeshire; &c.

Fig. 160. ASCLERA SANGUINICOLLIS, *Fab.* [*F.* Œdemeridæ. *G.* Asclera, *Schmidt.*] (Œdemera sanguinicollis, *Curtis.*)

Elongate, narrow, brassy green, with a very short fine yellowish-grey pubescence, thorax red. Head thickly and finely punctate, with a transverse impression between the antennæ, and a fovea on the crown; palpi black, reddish brown at the base: antennæ black, pubescent, two basal joints reddish brown beneath. Thorax truncate and transversely impressed in front and at the base; sides slightly rounded to about the middle, thence abruptly sinuate to the base; thickly and finely punctate, the disc with three large foveæ, one on each side a little before the middle, and one at the base in front of the scutellum. Elytra parallel at the sides, thickly and finely punctate, interstices rugulose; each with the suture and outer margin and three longitudinal lines on the disc, elevated, smooth. Length, 5–6 lines
In decayed trees, especially elms; Windsor and New Forests; Bristol; Richmond Park; &c.

Fig. 161. PYROCHROA COCCINEA, *L.*, *Curtis.* CARDINAL BEETLE. [*F.* Pyrochroidæ. *G.* Pyrochroa, *Geoffr.*]

Elongate, black, pubescent, thorax and elytra bright scarlet. Head triangular. with a transverse impression in front of the antennæ, and an elongate fovea on each side behind them, finely punctate, with a short black pubescence. Thorax transverse, narrowed anteriorly, rounded in front at the sides and at the base; its surface uneven, very finely and thickly punctate, scarlet, clothed with a depressed pubescence of the same colour; with a shallow, central, longitudinal furrow. Scutellum rounded at the apex, black, with a dense black velvety pubescence. Elytra dilated at the sides behind, bright scarlet, with a dense short pubescence. Underside and legs black. Length. 5–7 lines.
Not uncommon in Birch and Darenth Woods, Kent: the New Forest, and near Epping.
36

PLATE XIX.

Fig. 162. HYLECŒTUS DERMESTOIDES, *L.*, *Curtis*. (Male.) [*F.* Lymexylonidæ.
G. Hylecœtus, *Latr.*]

Elongate, black, rather shining, pubescent, legs reddish yellow, antennæ pitchy; elytra occasionally ferruginous, with the apex fuscous. Head transverse, nearly as wide as the thorax, thickly and rather coarsely punctate, with a large impression on the forehead; maxillary palpi with the terminal joint emitting a double row of elongate leaflets. Thorax transverse, slightly narrowed in front, sides nearly straight, rather thickly and somewhat coarsely punctate. Elytra elongate, sides parallel, very thickly and finely punctate, with traces of oblique longitudinal ridges. Length, 3–5½ lines.

Female, reddish brown; antennæ fuscous at the apex; terminal joint of maxillary palpi oblong, simple. Length, 4–8 lines.

On old birch-trees; Sherwood Forest; and Black Forest, Rannoch, Perthshire.

Fig. 163. LYMEXYLON NAVALE, *L.*, *Curtis*. (Female.) [*F.* Lymexylonidæ. G. Lymexylon, *Fab.*]

Elongate, slightly shining, clothed with a short grey pubescence, reddish yellow, head black, tips of antennæ, and outer margin and apex of elytra fuscous. Head transverse, wider than the thorax, sub-orbiculate, coarsely and very thickly punctate, sub-opaque; antennæ pitchy, basal joint rufous; terminal joint of maxillary palpi ovate, truncate, simple. Thorax longer than broad, narrowed in front, sides rounded anteriorly, nearly straight behind; minutely and thickly punctate. Elytra elongate, narrowed to the apex, gaping at the suture, shorter than the abdomen, finely and thickly punctate, each with two or three very indistinct longitudinal ridges. Length, 4–5½ lines.

Male black, elytra fuscous, the base within from the shoulder to the middle of the suture, abdomen, and legs, reddish yellow; maxillary palpi with the terminal joint emitting an intricate fascicle of hairy filaments. Length, 2½–4 lines.

Very rare. On old oaks, Windsor Forest; and on oak timber in Devonport Dockyard.

Fig. 164. ANTHICUS INSTABILIS, *Schmidt*. (Male.) [*F.* Anthicidæ. G. Anthicus, *Payk.*] (A. tibialis, *Curtis*, but not of *Waltl.*)

Oblong, pitchy black, with a yellowish-grey pubescence, thickly and rather coarsely punctate; antennæ, palpi, tibiæ, and tarsi reddish brown. Head as wide as the thorax, sub-quadrate. Thorax longer than broad, narrowed behind; sides rounded in front, nearly straight posteriorly; with a triangular fovea at the base in front of the scutellum. Elytra ovate, a little shorter than the abdomen. Posterior tibiæ dilated externally at the apex. (In the female the posterior tibiæ are simple.) Length, 1¼ lines.

Not uncommon. Netley; Ryde, Isle of Wight; Southampton; Southend; &c.

Fig. 165. GIBBIUM SCOTIAS, *Fab.*, *Curtis*. [*F.* Ptinidæ. G. Gibbium, *Scopoli.*]

Globose, shining chestnut brown; antennæ and legs clothed with a dense silky yellow pubescence. Head deflexed. Thorax transverse, narrowed in front, sides nearly straight. Elytra ovate, convex. Length, 1¼–1½ lines.

Found in old houses; Bristol; London; Newcastle-on-Tyne; &c.

Fig. 166. PTINUS SEXPUNCTATUS, *Panzer*, *Curtis*. [*F.* Ptinidæ. G. Ptinus, *L.*]

Sub-cylindrical; brown or reddish brown, forehead, scutellum, a large patch at the base of each elytron just behind the shoulder, and another, frequently divided, near the apex, snowy white;

PLATE XIX.—*Continued.*

antennæ and legs reddish brown, clothed with a short yellowish-grey pubescence. Thorax longer than broad, convex in front, depressed and constricted behind, with an acute tubercle on each side, and two oblong obtuse elevations between them, coarsely punctate, with a few long coarse yellowish grey hairs. Elytra with the sides nearly parallel, each with nine rows of oblong punctures; interstices transversely rugulose, with a few scattered punctures and a row of depressed yellowish grey hairs. Length, 2–2¼ lines.

In old oaks, and occasionally in houses: Edinburgh; Carlisle; London; Richmond Park; &c.

Fig. 167. MEZIUM AFFINE, *Boield.* [*F.* Ptinidæ. *G.* Mezium, *Curtis.*] (M. sulcatum, *Curtis*, but probably not of *Fabricius.*)

Globose, shining, chestnut brown, head, antennæ, thorax, and legs clothed with a dense ashy scale-like pubescence. Thorax a little broader than long, narrowed and margined behind; sides rounded in front, nearly straight posteriorly, with four longitudinal ridges, two entire and parallel in the middle, and one, abbreviated in front, on each side. Length, 1¼–1¾ lines.

Not uncommon in old houses in London.

Fig. 168. ANOBIUM PERTINAX, *L., Curtis.* [*F.* Anobiidæ. *G.* Anobium, *Fab.* Death-watches.]

Semi-cylindrical, pitchy brown or black, ruggedly punctate, with a dense short fine grey pubescence, opaque. Thorax transverse, narrowed behind, rounded in front and at the sides anteriorly, convex; depressed behind, with a deep oblong fovea in the middle and a broad one on each side, clothed with a dense pale yellow silky pubescence. Elytra elongate, sides parallel: punctate-striate, the punctures quadrate; interstices flat. Legs pitchy black, tibiæ and tarsi brown. Length, 2¼–2½ lines.

Very rare. In old oaks, near Bridgenorth, Shropshire.

Fig. 169. EUGLENES PYGMÆUS, *De Geer.* (Male.) [*F.* Pedilidæ. *G.* Euglenes, *Westwood.*] (Xylophilus oculatus, *Curtis.*)

Oblong, black, with a yellowish-grey pubescence, elytra reddish brown, antennæ and legs reddish yellow. Head wider than the thorax, transverse, punctate; eyes very large, approximate; antennæ as long as the body. Thorax transverse, narrowed in front, sides faintly rounded; thickly and finely punctate, with several obscure impressions at the base. Elytra with the sides nearly parallel, thickly and rather coarsely punctate.

In the female the eyes are smaller and distant, and the antennæ do not exceed half the length of the body. Length, 1–1½ lines.

Not common. In old hedges, and beneath bark; Colney Hatch; Coombe Wood; Epping Forest: &c.

Fig. 170. XYLETINUS ATER, *Panzer.* [*F.* Anobiidæ. *G.* Xyletinus, *Latr.*] (Serrocerus pectinatus, *Curtis.*)

Oblong, rather convex, black or pitchy, minutely granulate, slightly shining, sparsely clad with a short grey pubescence; base of antennæ and legs red. Head nutant. Thorax transverse, when viewed from above appearing narrowed in front, anterior angles deflexed, sides and posterior angles rounded and slightly reflexed, base bi-sinuate. Elytra finely punctate-striate, interstices rather convex. Length, 1½–2 lines.

In old trees; rare; North Mimms, Hertfordshire; Bridgenorth, Shropshire; Eastbourne, Sussex; Charlton, Kent; &c.

33

PLATE XX.

Fig. 171. ATTAGENUS VERBASCI, *L.* [*F.* Dermestidæ. *G.* Attagenus, *Latr.*]
(A. trifasciatus, *Curtis.*)

Oblong ovate, black, pubescent. Head small, deflexed, finely punctate, with long depressed silvery grey hairs. Thorax transverse, narrowed in front, rounded at the sides, base bi-sinuate, finely punctate, the sides and posterior margin covered with a long depressed yellowish-grey pubescence. Elytra brownish black, minutely punctate, with a small patch near the scutellum, one at the apex, and three transverse flexuous bands of yellowish-grey pubescence. Legs pitchy black; tibiæ and tarsi brown. Length, 1½–2 lines.

" In birds' skins, and houses, Edinburgh."

Fig. 172. NOSODENDRON FASCICULARE, *Oliv., Curtis.* [*F.* Byrrhidæ. *G.* Nosodendron, *Latr.*]

Ovate, convex, shining black. Head triangular, minutely and sparsely punctate; antennæ with the club reddish yellow, pilose. Thorax transverse, narrowed in front, sides slightly rounded and reflexed; base bi-sinuate; with a central longitudinal furrow, abbreviated in front, and a forea midway between it and the posterior angles; sparsely and rather coarsely punctate. Elytra deeply, coarsely, and thickly punctate, each with five rows of minute fascicles of short erect yellow setæ. Legs pitchy black. Length, 2½ lines.

Said to have been taken in Speechwick Park, near Ashburton, Devon; and at Southend, Essex.

Fig. 173. ASPIDIPHORUS ORBICULATUS, *Gyll., Curtis.*) [*F.* Sphindidæ. *G.* Aspidiphorus, *Latr.*]

Sub-orbiculate, pitchy black or black, finely pubescent, antennæ and legs reddish yellow, club of the former fuscous. Head transverse, very finely and thickly punctate. Thorax transverse, rather convex, narrowed in front, sides rounded anteriorly, base bi-sinuate; very finely and thickly punctate. Elytra much wider than the thorax, transverse, rounded at the sides, rather coarsely punctate-striate; interstices flat, smooth. Length, ¾–¾ line.

Not common. In moss and decaying wood. Colney Hatch; near Reigate; Hampstead; Epping Forest; &c.

Fig. 174. SIMPLOCARIA SEMISTRIATA, *Fab., Curtis.* [*F.* Byrrhidæ. *G.* Simplocaria, *Steph.*]

Ovate, convex, pitchy black or brown, with a brassy tint, clothed with a dense short erect grey pubescence; antennæ and legs reddish yellow. Head small, transverse, rather coarsely and sparsely punctate. Thorax transverse, narrowed in front, sides slightly rounded, base bi-sinuate, finely and rather thickly punctate; each with an entire sutural stria, and five striæ at the base, obliterated before the middle. Length, 1½ lines.

Common. On shady mossy banks in sandy districts.

PLATE XX.—*Continued.*

Fig. 175. TIRESIAS SERRA, *Fab.* [*F.* Dermestidæ. *G.* Tiresias, *Steph.*] (Megatoma serra, *Curtis.*)

Oblong ovate, black or pitchy black, sparsely clothed with a short grey pubescence: antennæ and legs reddish yellow. Head small, finely punctate. Thorax transverse, narrowed in front, rounded at the sides, base bi-sinuate, finely and sparsely punctate. Elytra oblong, rather coarsely and thickly punctate. Length, 1¾–2 lines.

Not common. On old trees and palings; Kensington Gardens; Battersea; Richmond Park; Hampstead; &c.

Fig. 176. LAMPROSOMA CONCOLOR, *Sturm.* [*F.* Lamprosomidæ. *G.* Lamprosoma, *Kirby.*] (Oomorphus concolor, *Curtis.*)

Ovate, convex, black, glossy, with a faint brassy tint: second joint of antennæ reddish yellow. Head very obscurely punctate. Thorax transverse, sides faintly rounded, base bi-sinuate : finely and sparsely punctate. Elytra with oblique rows of large rather remote punctures, the row next the suture short : interstices flat, with minute scattered punctures. Length, 1½ lines.

Not uncommon in many places in the south; Dover; Southampton; Isle of Wight; Hampstead; Finchley; &c.

Fig. 177. DERMESTES LARDARIUS, *L., Curtis.* BACON-BEETLE. [*F.* Dermestidæ. *G.* Dermestes, *L.*)

Elongate, sub-cylindrical, black, pubescent, opaque, elytra with a broad testaceous band at the base. Head finely and very thickly punctate, with a sparse yellow pubescence : antennæ reddish brown. Thorax transverse, narrowed in front, sides rounded, base bi-sinuate, thickly and rather finely punctate, with a depressed black pubescence, and numerous minute patches of yellowish-grey hairs. Scutellum with a black pubescence. Elytra thickly and finely punctate, each with a broad, dentate, testaceous, thickly cinereous pubescent band at the base, in which is a transverse row of three or four black spots. Length, 3–4 lines.

Common in larders and warehouses in which bacon and skins are stored.

Fig. 178. BYRRHIUS DENNII, *Curtis.* [*F.* Byrrhidæ. *G.* Byrrhus, *L.* PILL-BEETLES.]

Ob-ovate, convex, black, clothed with a short golden-yellow pubescence. Head finely and thickly punctate, with a faint transverse impression in front, and two minute foveæ behind; antennæ reddish brown. Thorax transverse, narrowed in front, sides waved, base bi-sinuate ; finely and rather thickly punctate, with a central longitudinal impressed line, a large round black spot on each side in front, and three or four more or less united black patches on each side behind. Elytra finely punctate-striate, with a transverse black patch in the middle, and each with five more or less interrupted longitudinal black stripes. Legs pitchy black. Length, 4 lines.

Rare. "In a chalk-pit at Barham, Suffolk ;" and occasionally in sand-pits on Hampstead Heath.

Fig. 179. THROSCUS OBTUSUS, *Curtis.* [*F.* Throscidæ. *G.* Throscus, *Latr.*]

Oblong ovate, ferruginous brown, slightly shining, clothed with a depressed yellowish-grey pubescence. Head convex, punctate. Thorax transverse, narrowed in front, rounded at the sides, base bi-sinuate, posterior angles acute, sparsely punctate. Elytra oblong ovate, finely striate, the striæ very finely punctate ; interstices flat, minutely punctate. Length, ⅔ line.

Very rare. On oaks: " Ensham, Oxon, and Plaistow Marshes, Essex."

40

PLATE XXI.

Fig. 180. ANTHAXIA NITIDULA, *L.* [*F.* Buprestidæ. *G.* Anthaxia, *Eschs.*] (Buprestis nitidula, *Curtis.*)

Elongate, golden green, shining. Head thickly, shallowly, rugulose-punctate, with a sparse, short grey pubescence; antennæ black, the basal articulations with a metallic green tint. Thorax transverse, narrowed behind; anterior margin bi-sinuate; sides rounded anteriorly, sinuate posteriorly; posterior angles nearly rectangular; with a shallow, longitudinal, impressed, central line, a small round fovea in the middle on each side of the central line, and a large, ill-defined impression near the hinder angles; thickly rugulose-punctate. Scutellum black, thickly and finely punctate. Elytra a little wider than the thorax, depressed, with the sides parallel three-fourths of their length, thence rounded to the apex; rather thickly and coarsely punctate, the interstices granulate. Legs black, femora and tibiæ with a green tint. Length, 2½ lines.
Rare. In flowers of white-thorn, New Forest, near Brockenhurst.

Fig. 181. APHANISTICUS PUSILLUS, *Olivier, Curtis.* [*F.* Buprestidæ. *G.* Aphanisticus, *Latr.*]

Elongate, brassy black, shining. Head large, convex, with a deep, broad, central, longitudinal furrow, and a few large, shallow, ocellate impressions. Thorax transverse; the sides flattened, lateral margins reflexed; base bi-sinuate, posterior angles acute; with two transverse impressions, one just before, the other a little behind the middle, and, usually, an elongate fovea in front of the scutellum; with a few scattered, large, shallow, ocellate impressions. Elytra at the base a trifle wider than the thorax, gradually widening at the sides to about two-thirds of their length, thence rather abruptly narrowed to the apex; coarsely rugulose-punctate, the punctures here and there in rows. Length, 1¼–1½ lines.
Not common. In moss, and by brushing in grassy places; Combe Wood; Southend; Bath; Farnham; &c.

Fig. 182. EROS MINUTUS, *Fab.* [*F.* Lycidæ. *G.* Eros, *Newman.*] (Lycus minutus, *Curtis.*)

Elongate, black, pubescent, opaque, elytra blood-red. Head triangular, with a deep fovea between the antennæ; eyes large, prominent; antennæ with the entire apical, and the anterior portion of the penultimate joints pale yellow. Thorax sub-quadrate, narrowed in front; the sides nearly straight; posterior angles acute; all the margins reflexed; with six foveæ, formed by two oblique, one longitudinal, and two transverse ridges, uniting a little behind the middle. Elytra with the sides parallel, each with the suture elevated and four longitudinal ridges, and between them two rows of quadrate punctures. Length, 2½–3 lines.
Very rare, although widely distributed in the south; Linton, Cambridgeshire; Tunbridge Wells; Coombe, Birch, and Darenth Woods; near Croydon; Bristol; Norfolk; Devon; &c.

Fig. 183. AGRILUS SINUATUS, *Olivier.* [*F.* Buprestidæ. *G.* Agrilus, *Solier.*] (Agrilus chryseis, *Curtis.*)

Elongate, above coppery purple, opaque. Head rugulose-punctate, with a central, longitudinal, impressed line; antennæ brassy black. Thorax transverse, narrowed behind; anterior margin bi-sinuate; sides rounded in front, sinuate posteriorly; base bi-sinuate, posterior angles acute; coarsely punctate, interstices obliquely rugulose; with a broad, shallow, longitudinal, central impression, and a deep, oblong fovea, bounded externally by a fine, curved, longitudinal ridge, on each side, near the posterior angles. Elytra at the base a little wider than the thorax, shoulders prominent, sides gradually widened to about two-thirds of their length, thence narrowed to the apex; separately rounded and finely serrated at the tips; each with a large, deep, circular impression at the base, and a broad, shallow, longitudinal furrow next the suture; coarsely granulate. Beneath brassy black, apical abdominal segment entire, pro-sternum deeply emarginate in front. Femora and tibiæ brassy black; tarsi black, claws bifid. Length, 3½–4½ lines.
Rare. New Forest; Windsor.

G   41

PLATE XXI.—*Continued.*

Fig. 184. MEGAPENTHES LUGENS, *Redtenb.* [*F.* Elateridæ. *G.* Megapenthes, *v.* Kiesew.] (Elater aterrimus, *Curtis*, but not of *Linné.*)

Elongate, black. opaque, sparsely pubescent. Head semicircular, convex, thickly and rather coarsely punctate ; antennæ serrated, longer than the head and thorax. Thorax elongate, convex, narrowed in front : sides slightly rounded anteriorly, straight behind ; posterior angles acute, with a short, very prominent ridge ; thickly and rather coarsely punctate, with an obscure, central, longitudinal, impressed line behind. Elytra at the base a trifle narrower than the thorax, gradually narrowed to the apex ; sides nearly straight : separately emarginate-truncate at the tips ; slightly shining, deeply crenulate-striate. interstices sparsely rugulose-punctate. Legs black, tips of femora, and occasionally the tarsi, ferruginous. Length, 4½–5 lines.

Very rare. Windsor Forest ; Tunbridge Wells ; Holloway, near London ; Sydenham.

Fig. 185. MELASIS BUPRESTOIDES, *L.*, *Curtis.* (Male.) [*F.* Melasidæ. *G.* Melasis, *Olivier.*]

Elongate, pitchy black, sub-opaque. Head large, narrower than the thorax, thickly and rather coarsely punctate, with three shallow, longitudinal impressions in front : antennæ pitchy. Thorax transverse, narrowed behind : sides nearly straight ; posterior angles acute, prominent : covered with minute asperities, and having a fine, central, longitudinal, impressed line. Elytra wider at the base than the thorax, sides nearly straight ; deeply striate, the striæ minutely punctate : interstices slightly convex, thickly and coarsely rugulose-punctate. Legs pitchy. Length, 3–4 lines.

Not common. Henhault, Windsor, and New Forests ; Hampstead ; &c. In old trees, especially oaks.

Fig. 186. TRACHYS MINUTUS, *L.*, *Curtis.* [*F.* Buprestidæ. *G.* Trachys, *F.*]

Elongate ovate, brassy black, with a faint violet tint. Head very finely reticulate, with a large, deep, triangular impression between the eyes ; clothed with a sparse, short grey pubescence. Thorax transverse, narrowed in front. sides nearly straight, with an impression on each side near the anterior angles, and an impressed, punctate, transverse line at the base ; very finely reticulate, and sparsely clothed with a short grey pubescence. Elytra at the base a little wider than the thorax, short, triangular, the sides rounded, shoulders prominent : each with a transverse impression at the base, a few irregularly-disposed, large, shallow punctures, and three transverse, flexuous bands of silvery-grey hairs. Length, 1¼–1½ lines.

Not uncommon on sallow, birch, and hazel, in many places in the south.

Fig. 187. DASCILLUS CERVINUS, *L.*, *Curtis.* (Female.) [*F.* Dascillidæ. *G.* Dascillus, *Latr.*]

Oblong ovate, convex, livid, clothed throughout with a dense, depressed yellowish-grey pubescence ; antennæ and legs testaceous. Head small, thickly and finely punctate, with an oblong tubercle on each side at the base of the antennæ. Thorax transverse, narrowed in front : the sides rounded behind and reflexed ; base bi-sinuate, posterior angles nearly rectangular ; finely and thickly punctate. Elytra at the base wider than the thorax, slightly dilated at the sides behind, finely punctate, the punctures arranged here and there in irregular rows : with several indistinct longitudinal ridges.

The males are black or fuscous, and the antennæ and legs, with the exception of the claw-joints, which are testaceous, of the same colour. Length, 4½–5 lines.

Common, and widely distributed.

Fig. 188. HYDROCYPHON DEFLEXICOLLIS, *Müller.* [*F.* Cyphonidæ. *G.* Hydrocyphon, *Redtenb.*] (Elodes-pini, *Curtis.*)

Ovate. convex, pitchy black, pubescent, shining. thickly and minutely punctate, parts of the mouth, base of antennæ, and legs yellow. Head transverse. Thorax transverse, narrowed and truncate in front, sides rounded, base bi-sinuate. Elytra very convex, clothed with a depressed yellow pubescence. Legs pale yellow, tarsi fuscous. Length, ¾–⅞ line.

Abundant in marshy places in the north.

PLATE XXII.

Fig. 189. TRICHODES ALVEARIUS, *Fab.* [*F.* Cleridæ. *G.* Trichodes, *Herbst.*] (Clerus alvearius, *Curtis.*)

Elongate, blue-black, clothed with long black and grey hairs intermixed, elytra bright red, with the suture, a spot round the scutellum, an oblique band one-third from the base, a transverse dentate band a little behind the middle, and a patch at the apex blue-black. Head and thorax thickly and coarsely punctate, the latter with a narrow, central, longitudinal, slightly raised space. Elytra thickly and coarsely punctate, the punctures arranged here and there in irregular rows. Beneath blue-black, with a dense long grey pubescence. Length, 5 7 lines.
Very rare. Near Dover; Manchester.

Fig. 190. LAMPYRIS NOCTILUCA, *L., Curtis.* (Male.) [*F.* Lampyridæ. *G.* Lampyris, *L.*]

Elongate, slightly shining, brownish-black, clothed with a short grey pubescence, thorax with the margins testaceous yellow. Thorax slightly transverse, faintly narrowed in front; sides rounded anteriorly, nearly straight behind; posterior angles nearly rectangular, extreme apices obtuse; thickly and rather coarsely punctate, with an obscure, central, longitudinal channel. Elytra with the sides nearly parallel, thickly and somewhat coarsely rugulose-punctate, each with three longitudinal, oblique, raised lines. Beneath fuscous, the margins of the segments narrowly testaceous. Legs fusco-testaceous, apex of tibiæ and tarsi dusky. Length, 6–8 lines.
Grassy banks, and margins of pathways in woods; widely distributed.

Fig. 191. LAMPYRIS NOCTILUCA, *L., Curtis.* (Female.) GLOW-WORM. [*F.* Lampyridæ. *G.* Lampyris, *L.*]

Elongate, reddish-brown, slightly shining, clothed with a short grey pubescence, margins of the thorax testaceous yellow. Thorax transverse, nearly semicircular, thickly and rather coarsely punctate, with a distinct central longitudinal channel. Elytra and wings entirely wanting. Abdomen thickly and coarsely punctate, with a longitudinal ridge in the centre, the lateral and anterior margins of the segments reddish-yellow, apical segment pale yellow; beneath reddish-brown, the three apical segments pale yellow. Legs fusco-testaceous, tips of tibiæ and tarsi dusky. Length, 6–9 lines.
Grassy banks, and margins of pathways in woods; generally distributed.

Fig. 191 *bis.* NECROBIA RUFICOLLIS, *Fab., Curtis.* [*F.* Cleridæ. *G.* Necrobia, *Latr.*]

Oblong, clothed with a yellowish-grey pubescence, shining, blue; thorax, scutellum, base of elytra, and legs red. Head transverse, rather coarsely and thickly punctate, with an elongate fovea on each side between the antennæ; parts of the mouth pitchy; antennæ black. Thorax transverse, narrowed in front; sides rounded; base bi-sinuate; posterior angles obtuse: thickly and coarsely punctate. Elytra at the base wider than the thorax, slightly dilated behind, each with eight rows of oblong punctures, interstices minutely and sparsely punctate. Beneath, breast red, abdomen black. Length, 2–2¾ lines.
Common on carcases, and in bone-yards.

Fig. 192. MALACHIUS MARGINELLUS, *Fab.* (Male.) [*F.* Malachiidæ. *G.* Malachius, *F.*] (M. bispinosus, *Curtis.*)

Elongate, green, sometimes with a blue or violet tint, shining, clothed with an exceedingly fine, short yellowish-grey pubescence, and long erect black hairs; front of head, sides of thorax, and apex of elytra, reddish-yellow. Head very finely and rather thickly punctate, with a broad transverse impression between the antennæ; anterior portion and parts of the mouth reddish-yellow, tips of mandibles and of palpi black; antennæ dusky, the basal joints with a green tint, the four first ferruginous beneath; basal joint robust; second triangular; 3–7 inclusive, with the apex produced into a curved tooth within. Thorax sub-orbiculate, rounded in front and at the sides, less so at the base; lateral margins reflexed, reddish-yellow; thickly and finely punctate. Elytra a trifle wider than the thorax; sides nearly straight; very thickly and minutely rugulose-punctate : the tips with a reddish-yellow patch within; each with the inner apical angle folded in; the upper edge of the hollow thus formed triangularly produced, with a black spine at the summit, and a deflexed, slightly curved, black, spiniform process beneath. Underside brassy green, sides of

43

PLATE XXII.—*Continued.*

prothorax, epimera of mesothorax, and apical margins of abdominal segments, reddish-yellow. Legs brassy green, pubescent, extreme tips of coxæ and of femora reddish-yellow, the anterior, with the apices of the tibiæ and the tarsi, reddish-yellow. In the female, the joints of the antennæ are feebly and gradually dilated within at the apex, the second ob-conic; the elytra entire and separately rounded at the apex. Length, 2½ lines.
Apparently a littoral species, occurring, not uncommonly, at Southend; Dover; Folkestone; and on the coast of Norfolk : Suffolk; Devon; &c.

Fig. 193. TELEPHORUS ABDOMINALIS, *Fab.* [*F.* Telephoridæ. *G.* Telephorus, *Schaeffer.*] (T. cyaneus, *Curtis.*)

Elongate ; head, antennæ, and legs black ; apex of mandibles and palpi pitchy ; mouth, basal joint of antennæ, thorax, anterior coxæ, and abdomen beneath reddish-yellow; elytra blue-black, rugulose-punctate, with a short sub-erect black pubescence. Length, 5–6½ lines.
Rare. In the north of England, and in Scotland.

Fig. 194. TARSOSTENUS UNIVITTATUS, *Rossi.* [*F.* Cleridæ. *G.* Tarsostenus, *Spinola.*] (Opilus fasciatus, *Curtis.*)

Elongate, pitchy black, shining, sparsely clothed with an erect yellowish-grey pubescence ; elytra with a slightly oblique, transverse, interrupted white band behind the middle. Head convex, coarsely and rather thickly punctate ; parts of the mouth and base of antennæ reddish-brown. Thorax elongate, truncate in front, narrowed behind; sides straight anteriorly, rounded posteriorly: hinder angles obtuse ; coarsely and thickly punctate at the sides ; with a longitudinal, mesial, coarsely-punctate impression, bounded on each side by a glabrous ridge in front, dilated posteriorly, and enclosing an oblong, raised, smooth space. Elytra elongate, sub-cylindrical, each with ten closely-set rows of large punctures, which gradually diminish in depth and regularity towards the apex, and an oblique, slightly arcuate, transverse white fascia a little behind the middle. Thighs pitchy black, their tips, tibiæ, and tarsi reddish-brown. Length, 2¼ lines.
A native of Southern Europe, of excessively rare occurrence in Britain, the only registered localities being, Kent ? ; Winchmore Hill ; and Malvern.

Fig. 195. THANASIMUS FORMICARIUS, *L., Curtis.* [*F.* Cleridæ. *G.* Thanasimus, *Latr.*]

Elongate. pubescent. slightly shining. Head black, as wide as the thorax, thickly and rather coarsely punctate ; antennæ and palpi pitchy. Thorax as broad as long, narrowed and constricted behind, truncate in front ; sides nearly straight anteriorly, rounded posteriorly ; base rounded ; rather thickly and coarsely punctate, with a shallow, central, longitudinal channel behind, divided in front a little before the middle, each branch curved and extending to the lateral margin, the anterior space black, the posterior portion red. Elytra oblong, wider than the thorax, black, with a depressed black pubescence, the base reddish-yellow : each with a narrow, transverse, flexuous band of coarse white hairs a little behind the red basal space, and a broad, nearly straight one near the apex ; with six irregular rows of large punctures at the base, but which do not extend beyond the basal white band. Abdomen beneath bright red. Legs black, with long erect white hairs ; tarsi pitchy red. Length, 2¼–4½ lines.
Not uncommon ; upon and beneath the bark of old trees tenanted by *Hylesini* and *Scolyti.*

Fig. 196. TILLUS UNIFASCIATUS, *Fab., Curtis.* [*F.* Cleridæ. *G.* Tillus, *Olivier.*]

Elongate. pubescent, shining, black ; elytra red at the base, with a transverse yellowish-white band, interrupted at the suture, towards the apex. Head black, as wide as the thorax, minutely and sparsely punctate, with erect black hairs. Thorax elongate, narrowed behind : sides nearly straight in front, sinuate behind : posterior angles nearly rectangular; black ; sparsely and finely punctate, with a faint, transverse impression one-third from the apex, and very long black hairs at the sides. Scutellum black. Elytra elongate, black, basal two-fifths red : each with a transverse, slightly-bent, yellowish-white band, which does not reach the suture, a little behind the middle : with a sparse, mingled grey and black pubescence, the apex with silvery-grey hairs ; and with nine rows of large punctures at the base, terminating abruptly at the transverse band. Underside and legs black. Length, 2½–3½ lines.
Rare. On old oaks and oak palings: Windsor; near Hertford: Barnet; and at Camberwell, Surrey.

PLATE XXIII.

Fig. 197. BOSTRICHUS CAPUCINUS, _L._ [_F._ Bostrichidæ. _G._ Bostrichus, _Geoff._]
(Apate capucinus, _Curtis._)

Elongate, black, slightly shining, elytra, and four apical segments of abdomen beneath, red. Head small, nutant, thickly and rather finely rugulose-punctate, with a fine, rather long, dusky pubescence; antennæ and parts of mouth pitchy. Thorax convex, rounded in front and at the sides anteriorly; posterior angles obtuse; coarsely granulate, depressed and tuberculate in front. Elytra coarsely and deeply punctate, each with two or three obscure longitudinal ridges on the disc. Length, 5 lines.

Very rare. Old trees; Norfolk; Northampton; Derbyshire; &c.

Fig. 198. NEMOSOMA ELONGATUM, _L._, _Curtis._ [_F._ Trogositidæ. _G._ Nemosoma, _Latr._]

Long, narrow, cylindrical, black; antennæ, legs, and two spots on the elytra reddish-yellow. Head of the width of, and as long as, the thorax, rather thickly punctate, the punctures elongate, with a broad, central, longitudinal furrow in front; antennæ 10-jointed. Thorax elongate, narrowed behind, sides slightly rounded, finely and rather thickly punctate. Elytra elongate, nearly cylindrical, slightly dilated at the sides behind, nearly thrice as long as the thorax; finely punctate, the punctures here and there arranged in indistinct rows, sutural stria deep at the apex, and continued round the tips; their basal third, and a large, common, more or less transverse patch near the apex, reddish-yellow. Beneath black, with the margins of the first four, and the entire fifth, abdominal segments reddish-brown. Length, 2½–2¾ lines.

Rare. Inhabits beneath the bark of elms infested by _Hylesinus vittatus_, in whose burrows it resides, preying on the larvæ of the _Hylesini_; it has occurred near Sydenham, and Darenth, Kent; and in the vicinity of Nottingham.

Fig. 199. PLATYPUS CYLINDRUS, _Fab._, _Curtis._ (Male.) [_F._ Platypidæ. _G._ Platypus, _Herbst._]

Elongate, sub-cylindrical, sparsely pubescent, black; elytra reddish-brown, fuscous towards the apex. Head nearly as wide as the thorax, flat, opaque, and rugulose-punctate in front, glossy and coarsely punctate behind; antennæ reddish-yellow. Thorax elongate, sub-cylindrical, slightly narrowed posteriorly; sides nearly straight, sinuate a little behind the middle; base bi-sinuate; finely and rather thickly punctate; with an anteriorly abbreviated, central, longitudinal impressed line. Elytra elongate, cylindrical; each with eight longitudinal punctate furrows; interstices convex, smooth; retuse, densely pubescent, and bi-dentate at the apex, the inner denticulation small, the outer one acute, divergent. Legs reddish-brown, tarsi testaceous, anterior thighs bi-dentate. Beneath pitchy black, finely and rather sparsely punctate and pubescent. In the female, the elytra are entire, and unarmed at the apex. Length, 3⅓–3¾ lines.

Rare. In old oaks; Windsor and New Forests.

Fig. 200. HYLESINUS OLEIPERDA, _Fab._ (Male.) [_F._ Hylesinidæ. _G._ Hylesinus, _Fab._] (H. scaber, _Curtis._)

Broad, ovate, convex, black, clothed with a depressed black and fulvous pubescence. Head thickly and finely rugulose-punctate, with a semicircular, deeply-impressed pubescent space in front, between the antennæ; antennæ reddish-yellow, club fuscous, its apex pale. Thorax transverse, convex, narrowed in front; sides rounded, base bi-sinuate; thickly rugulose-punctate, with a faint, central, longitudinal furrow, and a transverse arcuate impression near the posterior margin. Elytra oblong, convex; each with nine deep longitudinal striæ; interstices convex, granulate, that next the suture with a dense fulvous pubescence. Legs pitchy black, tarsi reddish-yellow. In the female the forehead is flat. Length, 1–1½ line.

Not common. Beneath bark of ash and lilac; Leicestershire; Mickleham and Reigate, Surrey; &c.

45

PLATE XXIII.—*Continued.*

Fig. 201. SCOLYTUS DESTRUCTOR, *Olivier, Curtis.* [*F.* Scolytidæ. *G.* Scolytus, *Geoffr.*]

Oblong, convex, pitchy black, shining; parts of mouth, antennæ, anterior margin of thorax, elytra, and legs reddish-yellow. Head flat in front, finely rugulose-punctate, with a large, dense patch of fulvous hairs. Thorax transverse, narrowed in front, sparsely and deeply punctate at the sides, faintly on the disc. Scutellum sunk, opaque. Elytra oblong, nearly quadrate, reddish-yellow or chestnut, sometimes with a fuscous patch or transverse band in the centre; each with six deeply punctate striæ; interstices flat, each with an irregular row of minute punctures, that next the suture wide, depressed at the base, and roughly punctate. Beneath pitchy black; the third and fourth abdominal segments with a small tubercle in the centre, on the anterior margin. Length, 1½–3 lines.

Common in the south. Beneath the bark of the limbs and trunks of recently felled and sickly elms.

Fig. 202. BARIDIUS ANALIS, *Oliv.* [*F.* Cryptorhynchidæ. *G.* Baridius, *Schoenh.*] (Baris analis, *Curtis.*)

Oblong, black, shining, apical third of elytra red. Head and rostrum minutely and sparsely punctate, the latter curved. Thorax elongate, narrowed in front, sides rounded anteriorly, nearly straight behind; thickly and rather coarsely punctate, with a narrow, smooth, longitudinal, central space in front. Elytra deeply and broadly striate, the striæ minutely punctate; interstices flat, each with a row of minute punctures. Length, 2 lines.

Very rare. Near Ryde, Isle of Wight.

Fig. 203. CIS BIDENTATUS, *Oliv., Curtis.* (Male.) [*F.* Cissidæ. *G.* Cis, *Latr.*]

Oblong, black, brown, or yellow, slightly shining, sparsely clothed with a short, sub-erect yellowish-grey pubescence. Head concave in the middle, forehead flat, anterior margin reflexed, bi-dentate; antennæ and parts of mouth reddish-yellow. Thorax transverse, convex; sides gently rounded; anterior angles acute, prominent; anterior margin produced over the head, bi-tuberculate; finely and rather thickly punctate. Elytra barely twice as long as the thorax, always of a lighter colour than it, rather less thickly and finely punctate. Beneath pitchy black, obscurely punctate. Legs reddish-yellow. In the female, the anterior margin of the head is but slightly reflexed, its angles rounded, and the produced anterior margin of the thorax is rounded and destitute of tubercles. Length, 1–1½ lines.

Not uncommon in boleti on ash.

Fig. 204. CICONES VARIEGATUS, *Hellwig.* [*F.* Synchitidæ. *G.* Cicones, *Curtis.*] (C. carpini, *Curtis.*)

Oblong, above black or pitchy, sub-opaque, clothed with a sparse, short, sub-erect brown and yellow pubescence, elytra with irregular yellow bands or spots. Head small, with a slightly-curved, impressed, longitudinal line in front on each side between the antennæ, parts of mouth and antennæ ferruginous. Thorax transverse, slightly narrowed behind; sides rounded, margins reflexed and finely crenulate: posterior angles nearly rectangular; base bi-sinuate; disc convex. Elytra a little wider than the thorax, punctate-striate, alternate interstices raised, with numerous reddish-yellow patches, more or less united, and usually forming three interrupted, flexuous, transverse bands. Underside and legs ferruginous. Length, 1½–1¾ lines.

Rare. Beneath bark of beech and hornbeam; Epping and New Forests; and near Bromley, Kent.

Fig. 205. CORYNETES CŒRULEUS, *De Geer.* [*F.* Cleridæ. *G.* Corynetes, *Herbst.*] (C. violaceus, *Curtis.*)

Oblong, blue, or greenish blue, shining, clothed with long, erect black hairs. Head nearly as wide as the thorax, sparsely and rather finely punctate, antennæ and palpi black. Thorax nearly as broad as long, rounded at the sides, sparsely and rather coarsely punctate. Elytra wider than the thorax, with oblong punctures arranged in irregular rows at the base, interstices with a few minute punctures. Underside and legs black. Length, 2–2½ lines.

Not uncommon in the south.

46

PLATE XXIV.

Fig. 206. HYLURGUS PINIPERDA, *L., Curtis.* [*F.* Hylesinidæ. *G.* Hylurgus, *Latr.*]

Elongate, sub-cylindrical, pitchy black, shining, clothed with a very short erect grey pubescence; antennæ and tarsi ferruginous. Head convex, finely and thickly punctate at the sides, rather coarsely and sparsely in the middle; with an acute, smooth, central, longitudinal ridge in front. Thorax a trifle longer than broad, narrowed in front, slightly constricted at the apex, sides gently rounded, posterior angles rounded; rather coarsely and somewhat thickly punctate; with a narrow, slightly raised, smooth, central, longitudinal space. Scutellum thickly and finely punctate. Elytra a little wider than the thorax, raised and serrated at the base; each with eight rather coarsely punctate striæ; interstices flat, transversely rugulose, with a row of minute pubescent tubercles, reaching the extreme apex on the 1st and 3rd, but on the 2nd terminating at the commencement of the oblique portion. Length, 1½–1¾ lines.

Not uncommon. In and upon the young shoots of the Scotch fir (*Pinus sylvestris*).

Fig. 207. HYLURGUS PINIPERDA, *L., Curtis, var. d.* [*F.* Hylesinidæ. *G.* Hylurgus, *Latr.*]

The variety here represented is entirely of a pale reddish-yellow colour, and is probably merely an immature state; it occurs, not unfrequently, associated with the darker form.

Fig. 208. MESITES TARDII, *Curtis.* (Male.) [*F.* Cossonidæ. *G.* Mesites, *Schoenh.*] (Cossonus Tardii, *Curtis.*)

Elongate, pitchy black, sub-opaque; antennæ reddish-brown. Head rather thickly, coarsely, and deeply punctate in front, shallowly and sparsely behind, with an oblong fovea between the eyes; rostrum dilated before the apex, coarsely and rather sparsely punctate, with a deep, central, abbreviated, longitudinal furrow; antennæ inserted two-fifths from the apex. Thorax elongate, narrowed in front, constricted at the apex, sides rounded; thickly, coarsely, and deeply punctate, with a central, longitudinal, slightly raised, narrow, smooth space, terminating behind in a depression. Elytra a little wider than the base of the thorax, and about one-fourth longer, sides nearly straight; each with nine profound, deeply punctate, longitudinal striæ; interstices flat, finely rugulose-punctate. In the female, the antennæ are inserted at one-fifth from the base, and the portion of the rostrum in front of them is cylindrical, smooth, and of a reddish-brown colour. Length, 3–5¾ lines.

Local, but abundant where it occurs. Mountains of Wicklow and Kerry, in hollies; coast of Devonshire and Cornwall, in beech and sycamore; and, it is said, in Norfolk.

Fig. 209. MIARUS GRAMINIS, *Schoenh.* [*F.* Cionidæ. *G.* Miarus, *Schoenh.*] (Gymnetron graminis, *Curtis.*)

Short, ovate, black, clothed above with a fine, yellowish-grey, depressed pubescence. Head globose, rather coarsely punctate; eyes large; rostrum elongate, slender, cylindrical, slightly curved, obscurely punctate at the base, smooth from the base of the antennæ to the apex. Thorax transverse, convex narrowed in front; sides rounded; thickly, deeply, and rather coarsely punctate, with a faint, central, longitudinal ridge. Elytra broad, with profound, deeply and remotely punctate, longitudinal striæ; interstices broader than the striæ, flat, rugulose, each with two rows of yellowish-grey hairs. Length, 1½–1¾ lines.

Not uncommon. Amongst herbage on hedge-banks; near London; Cambridge; Hertford; &c.

Fig. 210. ORCHESTES PRATENSIS, *Germ.* [*F.* Erirhinidæ. *G.* Orchestes, *Illiger.*] (O. Waltoni, *Curtis.*)

Elongate, ovate, black, clothed with a short yellowish-grey pubescence; antennæ and tarsi reddish-brown, club of the former dusky. Head thickly and finely punctate; rostrum slender.

47

PLATE XXIV.—*Continued.*

Thorax transverse, narrowed in front, sides rounded, thickly and finely punctate. Elytra a trifle wider at the base than the thorax, oblong ovate, with deep, longitudinal, faintly punctate striæ; interstices minutely punctate. Posterior thighs with an angular projection beneath. Length, 1¼ lines. Not common. Knaresborough, Yorkshire ; near Hertford ; and near London.

Fig. 211. CŒLIODES GERANII, *Payk.* [*F.* Cryptorhynchidæ. *G.* Cœliodes, *Schoenh.*] (Ceutorhynchus Geranii, *Curtis.*)

Broad, ovate, convex, black, shining, sparsely clothed with a short, sub-erect black pubescence. Head and rostrum thickly and rather coarsely punctate, the latter deflexed, and slightly curved. Thorax convex, transverse, narrowed in front; sides rounded; anterior margin reflexed; thickly and rather coarsely punctate, with a transverse furrow in front, and a minute glabrous tubercle on each side. Elytra broad, wider at the base than the thorax; deeply striate, the striæ profoundly punctate; interstices flat, each with a row of small tubercles. Beneath with ashy scales. All the thighs with an exceedingly minute tooth on the underside. Length, 1½–1¾ lines. Not rare. On the meadow crane's-bill (*Geranium pratense*).

Fig. 212. ACALLES ROBORIS, *Curtis.* [*F.* Cryptorhynchidæ. *G.* Acalles, *Schoenh.*]

Ovate, convex, pitchy black, clothed with yellowish-grey, brown, and black scales: antennæ and rostrum reddish-brown, the latter sub-cylindrical, slightly bent, shining. minutely punctate. Thorax longer than broad, contracted in front, thence slightly narrowed to the base, with a central longitudinal channel obliterated in front, and an oblong tubercle on each side of it. Elytra broad, wider than the thorax at the base ; deeply striate : interstices convex, with patches of black scales forming irregular, transverse bands. Legs pitchy, clothed with ashy and brown scales. Length, 1¾–2 lines. Rare. But widely distributed in the south; on the oak.

Fig. 213. PHYTOBIUS COMARI, *Herbst.* [*F.* Erirhinidæ. *G.* Phytobius, *Schoenh.*] (Pachyrhinus comari, *Curtis.*)

Broad, ovate, rather convex, black: sides of thorax and underside clothed with white scales; elytra maculated with ashy scales. Head and rostrum thickly and rather coarsely punctate, with a few dispersed ashy scales: antennæ ferruginous, club pitchy. Thorax transverse, contracted in front ; sides rounded ; thickly and rather coarsely punctate, with a broad, shallow, central, longitudinal channel, and an obtuse tubercle on each side. Elytra at the base half as wide again as the thorax ; each with seven deep, broad, obscurely punctate, longitudinal striæ : interstices convex, rugulose-punctate. Thighs pitchy ; tibiæ and tarsi ferruginous ; claws fuscous. Length, 1¼–1½ lines. Not uncommon. In boggy places in the south, on the marsh cinquefoil (*Comarum palustre*).

Fig. 214. MONONYCHUS PSEUDACORI, *Fab., Curtis.* [*F.* Cryptorhynchidæ. *G.* Mononychus, *Germar.*]

Broad, ovate, rather depressed, black, sub-opaque, underside with ochre-yellow scales; elytra with a patch of yellowish-grey scales on the suture, at about one-fifth from the base. Head thickly and coarsely rugulose-punctate, with a depression in front between the eyes ; rostrum sparsely and rather finely punctate. Thorax transverse, very much narrowed in front: sides rounded, thickly and coarsely rugulose-punctate; with a central, longitudinal furrow, narrowed behind, and reaching the base, dilated and abbreviated in front. Elytra sub-quadrate, suture depressed at the base; punctate-striate: interstices flat, coarsely rugulose-punctate. Length, 2 lines. Abundant near Ventnor, Isle of Wight, in the seed capsules of the stinking iris, or roast-beef plant (*Iris fœtidissima*).

48

PLATE XXV.

Fig. 215. MAGDALINUS CARBONARIUS, *L.* [*F.* Erirhinidæ. *G.* Magdalinus, *Schoenh.*] (Magdalis carbonarius, *Curtis.*)

Elongate, black, moderately shining. Head minutely punctate, with a shallow impression between the eyes; rostrum a trifle longer than the thorax, slightly curved, thickly minutely rugulose-punctate; antennæ pitchy at the base. Thorax scarcely longer than broad, constricted in front; sides more or less rounded and armed anteriorly with three or four obtuse crenulations; thickly and coarsely punctate; with an obscure central raised line, abbreviated in front. Elytra shining, each with nine deep, broad, coarsely and closely punctate striæ; interstices convex, narrow, transversely strigose. All the femora with a minute tooth in the centre beneath. Length, 2¼–3½ lines. Rare. In the north of England, and in Scotland, in birch.

Fig. 216. OTIORHYNCHUS MAURUS, *G., C.* [*F.* Otiorhynchidæ. *G.* Otiorhynchus, *G.*]

Ovate, convex, black, clothed with a short, coarse, sparse pubescence, sub-opaque. Head and rostrum coarsely punctate, the latter with a central ridge; antennæ pitchy. Thorax transverse, convex, sides rounded, thickly and coarsely granulate, with an obscure central channel. Elytra ovate, faintly punctate-striate; interstices slightly convex, minutely granulate. Legs pitchy. Femora unarmed. Length, 3½–4 lines. Rare in the south; more frequent in the Shetland Isles.

Fig. 217. LIXUS ANGUSTATUS, *Fab., Curtis.* [*F.* Erirhinidæ. *G.* Lixus, *Fab.*]

Elongate, black, with a sparse, short, grey pubescence, and more or less densely covered with a yellow caducious pollinosity. Head and rostrum thickly and finely punctate; the former with a minute fovea between the eyes; the latter stout, sub-cylindrical, as long as the thorax, slightly curved, with a short central furrow between the antennæ. Thorax elongate, narrowed in front, sides slightly rounded; coarsely rugulose-punctate; with central furrow at base. Elytra nearly half as wide again as the thorax, and thrice as long, sides nearly parallel, thickly rugulose-punctate; each with ten rows of oblong, shallow punctures; interstices flat. Length, 6–9 lines. Not common. Sydenham, Kent; and on the Sussex coast, near Hastiugs, on thistles.

Fig. 218. ANTHONOMUS POMORUM, *L., Curtis.* APPLE-WEEVIL. [*F.* Erirhinidæ. *G.* Anthonomus, *Germar.*]

Elongate, pitchy brown, clothed with a fine, depressed, yellowish grey pubescence; elytra with a reddish yellow curved band behind. Head pitchy black, thickly punctate; rostrum black, longer than the head and thorax, slender, bent, with four deep furrows posteriorly; minutely and irregularly punctate anteriorly; antennæ ferruginous, club fuscous. Thorax transverse, narrowed in front, constricted at the apex; sides slightly rounded in front, nearly straight behind; disc rather flat; thickly and somewhat coarsely punctate. Scutellum clad with a dense white pubescence. Elytra oblong, wider than the thorax, slightly dilated at the sides posteriorly; each with ten coarsely punctate striæ; interstices flat, sparsely, and very minutely punctate; and with an oblique curved reddish yellow band, tessellated on its margins with small patches of white hairs, and bounded both in front and behind by a pitchy black bar. Thighs reddish brown, ferruginous at the tips the anterior armed beneath with a large, the posterior with a minute tooth; tibiæ and tarsi ferruginous. Length, 2–2½ lines. Common in orchards—in the spring on the blossoms; in the winter concealed in the crevices and beneath the scales of loose bark of the apple and pear-tree.

Fig. 219. ERIRHINUS ÆTHIOPS, *Fab., Curt.* [*F.* Erirhinidæ. *G.* Erirhinus, *Sch.*]

Elongate ovate, black, shining; antennæ and legs red brown, the club of the former fuscous. Head thickly punctate; rostrum as long as the head and thorax, bent, sparsely punctate, the punctures linear. Thorax nearly as long as broad, slightly rounded at the sides, straightly truncate at the base; coarsely punctate on the disc, more finely and thickly at the sides; with a narrow, smooth, central line. Elytra elongate ovate, thrice as long as the thorax; deeply punctate-striate; interstices flat, exceedingly finely punctate. Thighs unarmed. Length, 3 lines. Very rare. Found in marshy places at Askham Bryant, Yorkshire, and in Scotland.

H   49

PLATE XXV.—*Continued.*

Fig. 220. PHYTONOMUS FASCICULATUS, *Herbst.* [*F.* Cleonidæ. *G.* Phytonomus, Schoenh.] (Hypera fasciculosa, *Curtis.*)

Short ovate, black, clothed with black, reddish brown, yellow, and white scales. Head with reddish brown scales, its vertex and base of rostrum with ashy hairs; rostrum short, stout, deflexed, and slightly curved; antennæ ferruginous, pubescent. Thorax transverse, sides dilated and rounded, clothed with brown and yellowish scales, with a narrow central and broad lateral line of yellowish white scales. Elytra broad, covered with reddish brown scales; deeply striate; interstices convex, the sutural, third, and fifth broader than the others, and raised, clothed with whitish, and tessellated with black or brown scales. Legs reddish brown; thighs variegated with white, brown, and black scales; tibiæ and tarsi with coarse cinereous hairs. Length, 3–4 lines.

Not common, but very widely distributed. Dover; Deal; Norfolk; Isle of Thanet, Scotland, &c.

Fig. 221. RHYNCHITES OPHTHALMICUS, *Steph.* (Male.) [*F.* Attelabidæ. *G.* Rhynchites, *Herbst.*] (R. similis, *Curtis.*)

Oblong, deep blue, shining, with a fine, long, erect, black pubescence. Head narrow behind, widening to the eyes, sparsely and rather coarsely punctate; forehead flat, or slightly excavated; rostrum barely as long as the head, dilated towards the apex, sparsely punctate at the base, thickly at the tip; parts of the mouth pitchy; antennæ black. Thorax elongate, narrowed in front; sides slightly rounded anteriorly, sparsely and coarsely punctate. Elytra oblong, nearly twice as wide as the thorax, sides nearly parallel; striate; the striæ deeply, remotely, and coarsely punctate; interstices flat, each with two irregular rows of minute punctures. Legs greenish, tarsi black. The female has the head shorter, swollen behind the eyes, and the forehead convex. Length, 2¼–2½ lines.

Not uncommon in woods, in the south, on the whitethorn.

Fig. 222. APION DIFFORME, *Germ., Curt.* (Male.) [*F.* Attelabidæ. *G.* Apion, *Herbst.*]

Elongate ovate, black, the elytra faintly olivaceous; base of antennæ, of femora, and of tibiæ, reddish yellow. Head coarsely confluently punctate; rostrum nearly as long as the head and thorax, slightly bent, broad, with parallel sides to the base of antennæ, thence narrowed to tip, obscurely punctate; antennæ mesial, reddish yellow, apical articulations fuscous, basal joint elongate pear-shaped, 2d minute, 3d nearly as broad as long, compressed. 4th as wide as the 3d, but more elongate, 5th, 6th, 7th, and 8th, narrow, of equal width, the remaining three forming a slender elongate club. Thorax elongate, narrowed in front; sides rounded; thickly punctate, a central line behind. Elytra elongate ovate, much narrowed at apex, convex, deeply striate, the striæ finely punctate; interstices convex, smooth. Anterior tibiæ bent, basal joint of their tarsi elongate, with the apex produced on the inner side; posterior tibiæ robust, bent, the basal joint of their tarsi broad and flat. In the female the rostrum is longer, slender, cylindrical, and much curved; the antennæ are slender and black; the anterior tibiæ straight, their tarsi simple; the posterior tibiæ slender and straight, and the basal joint of tarsi does not exceed the others in width; tibiæ and tarsi entirely black; thighs reddish yellow, tips black. Length, 1¼ lines.

In grassy places in woods, in the autumn.

Fig. 223. POLYDROSUS SERICEUS, *Schall.* [*F.* Brachyderidæ. *G.* Polydrosus, Germar.] (P. speciosus, *Curtis.*)

Elongate, black, clothed with round, green, or bluish green scales; antennæ and legs pale reddish yellow, club of the former dusky. Head with a fovea in front; rostrum short, deflexed. Thorax transverse, narrowed in front, sides rounded. Elytra sub-cylindrical, each with ten profound coarsely and deeply punctate striæ; interstices flat. Femora and base of tibiæ with a few scattered green scales. Length, 3–4 lines.

Very rare. Kimpton, Hants.

PLATE XXVI.

Fig. 224. LÆMOPHLÆUS ATER, *Oliv.* [*F.* Cucujidæ. *G.* Læmophlæus, *Eric.*]
(Cucujus Spartii, *Curtis.*)

Elongate, parallel, flat, pitchy black or reddish brown, shining. Head finely and rather sparsely punctate, with a shallow elongate impression on each side in front between the antennæ. Thorax nearly as broad as long, narrowed behind, sides slightly rounded, posterior angles obtuse; very minutely punctate, with a fine longitudinal impressed line on each side. Elytra parallel, flat, each depressed in the middle, and with six striæ, of which the 2d, 3d, 4th, and 5th, are approximate, the narrow interstices between them raised, and contiguous to and outside the 6th a fine similarly raised line. Legs ferruginous, thighs clavate. Length, 1–1½ lines.
Rare. Beneath bark of broom, Coombe Wood, Surrey.

Fig. 225. SALPINGUS FOVEOLATUS, *Ljungh.* [*F.* Salpingidæ. *G.* Salpingus, *Illig*]
(Sphæriestes foveolatus, *Curtis.*)

Elongate, brassy black, smooth, shining; parts of mouth, base of antennæ, tibiæ and tarsi ferruginous. Head broad, sparsely and rather coarsely punctate. Thorax transverse, narrowed behind, sides rounded, rather thickly and coarsely punctate, uneven, with an arcuate impression on each side. Elytra punctate-striate at the base, irregularly punctate towards the apex, each with a large transverse, ovate, impression a little before the middle. Length, 1¾–2¼ lines.
Rare. North of England; Cramond, near Edinburgh.

Fig. 226. TROGOSITA MAURITANICA, *L.*, *Curtis.* [*F.* Trogositidæ. *G.* Trogosita, *Oliv.*]

Elongate, flat, pitchy black or reddish brown, shining. Head transverse, broad behind, narrowed anteriorly, forehead flat, rather sparsely, deeply, and somewhat coarsely punctate ; antennæ gradually and slightly thickened towards the apex. Thorax transverse, barely as wide as the elytra, narrowed behind; sides nearly straight anteriorly, slightly sinuate posteriorly ; anterior angles salient; posterior angles nearly rectangular, prominent; rather thickly, deeply, and coarsely punctate. Elytra slightly dilated at the sides posteriorly, humeral angles prominent, punctate-striate ; interstices flat, each with two rows of minute punctures. Legs ferruginous. Length, 3½–5 lines.
Not uncommon in bakehouses, houses, warehouses, and amongst ship biscuit; and occasionally beneath bark of dead trees.

Fig. 227. BRUCHUS CISTI, *F.* [*F.* Bruchidæ. *G.* Bruchus, *L.*] (B. ater, *Curtis.*)

Ovate, convex, black, clothed with a fine dense grey pubescence, shining. Head thickly and finely punctate ; antennæ black, the 2d, 3d, and 4th joints minute. Thorax transverse, narrowed in front, sub-conical, base with two oblique impressions ; finely and thickly granulate. Elytra profoundly striate, the striæ distinctly punctate; interstices flat, thickly and minutely punctate. Posterior thighs unarmed. Length, 1¼ lines. In the south, on the dwarf cistus.

Fig. 228. AROMIA MOSCHATA, *L.*, *Curtis.* MUSK-BEETLE. [*F.* Cerambycidæ. *G.* Aromia, *Serville.*]

Elongate, bright or bluish green, with a coppery or golden tint, shining. Head thickly and coarsely punctate, with a fine central impressed line behind; antennæ of male half as long again, of female scarcely as long as the body, dark blue, black towards the apex. Thorax transverse, armed on each side with an acute tubercle, uneven, transversely wrinkled in front and behind, coarsely punctate in the middle. Elytra wider than the thorax, thickly shagreened, each with two slightly raised longitudinal lines. Beneath brassy or coppery green, clothed with a dense white pile. Femora green with a slight blue tint, tibiæ and tarsi dark blue or violet. Length, 12–15 lines.
Locally abundant in the south, on old willows.

PLATE XXVI.—*Continued.*

Fig. 229. ÀTTELABUS CURCULIONOIDES, *L.*, *C.* [*F.* Attelabidæ. *G.* Attelabus, *L.*]

Broad ovate, convex, black, shining, thorax and elytra red. Head and rostrum coarsely punctate, the former with a deep linear impression on each side between the eyes; antennæ with the 2d, 3d, and 4th joints pitchy red. Thorax convex, transverse, narrowed in front; sides rounded; finely punctate. Scutellum black, with a few minute punctures. Elytra sub-quadrate, convex, each with eight shallow flexuous coarsely punctate striæ; interstices flat, sparsely and finely punctate. Underside and legs black. Length, 2–3 lines. Common in woods on young oaks.

Fig. 230. ANTHRIBUS ALBINUS, *L.*, *C.* (Male.) [*F.* Anthribidæ. *G.* Anthribus, *Geoffr.*]

Elongate, sub-cylindrical, thickly clothed with a reddish brown, grey, and black pile. Head and rostrum with a dense white pile, the former with a fine central line on the crown, and a transverse arcuate impression between the eyes, the latter with a triangular notch at the apex; antennæ black, with the tips of joints 1–7, the apical half of the 8th, and the basal half of the 9th, white. Thorax transverse, narrowed in front; sides rounded anteriorly, straight behind; the disc with three oblong tubercles surmounted by a minute dense fascicle of black bristles, the central one sometimes divided. Elytra wider than the thorax; convex clothed with a dense reddish brown pile; punctate-striate; the third interstice with a row of four or five small fascicles of black bristles; each with a more or less distinct irrorated white patch in the middle, and a broad white band occupying the apical fourth, with the exception of a small quadrate space at the inner angle. Beneath with a yellowish white pile. Femora reddish brown; tibiæ with a brown ring near the base, and another towards the middle, the apical third with a white pile; tarsi black with a few white hairs at the tip of each joint, claw joint covered at the base with white hairs. In the female the antennæ are black, the 8th joint alone covered with white pile. Length, 3½–5 lines. In the south; in old hedges, &c.

Fig. 231. PRIONUS CORIARIUS, *L.*, *Curt.* (Male.) [*F.* Prionidæ. *G.* Prionus, *Geoffr.*]

Pitchy black, shining. Head narrower than the thorax, coarsely punctate, with a central line; clypeus and palpi reddish brown, antennæ nearly as long as the body, 12-jointed, robust, tapering to the apex, the inner apical angle of each articulation very acute. Thorax transverse, with a fringe of yellow hairs in front, armed on each side with three large teeth, of which the central is the most prominent; base bi-sinuate with a cilia of yellow hairs; thickly rugulose-punctate at the sides, more sparsely on the disc. Scutellum coarsely punctate. Elytra wider than the thorax, sides nearly parallel, internal apical angles with a minute tooth; very coarsely and confluently punctate; each with three indistinct ridges. Beneath reddish brown; meso- and meta-thorax clothed with a yellowish grey down; abdomen sparsely and finely punctate, with a few short yellowish grey hairs. Legs pitchy black, tarsi brown, femora and tibiæ compressed. In the female the antennæ are shorter, slender, serrate, 11-jointed, the 11th and 12th joints soldered together. Length, 9–20 lines. Widely distributed in old oaks.

Fig. 232. PLATYRHINUS LATIROSTRIS, *F.*, *C.* [*F.* Anthribidæ. *G.* Platyrhinus, *Cl.*]

Elongate, depressed, thickly clothed with short black, brown, and grey hairs. Head broad, forehead flat, with a central line terminating anteriorly in a fovea; clothed in front with ochreous pubescence; rostrum broad, with two ridges in the centre, clothed with ochreous pubescence; mandibles and antennæ black. Thorax transverse, produced at the sides in the middle into a slightly emarginate lobe; depressed centre limited in front by an arcuate ridge; surface uneven; with two obscure transverse flexuous ridges, one in the middle, the other near the base. Elytra wider than the thorax, sides nearly parallel; each with ten rows of remote, minute, deep punctures; interstices flat, the 3d, 5th, and 7th, raised; apical fifth covered with a dense ochreous pile, and with three reddish yellow interrupted transverse bands. Legs black; thighs with a few grey hairs; tibiæ annulated with ochreous and reddish brown hairs; tarsi with ashy grey hairs at the base of each joint. Length, 5–6 lines.

Rare. Inhabits the boletus of the ash (*Sphæria fraxinea*); Norfolk; Bristol; Cheltenham; New Forest. &c.

PLATE XXVII.

Fig. 233. CLYTUS QUADRIPUNCTATUS, *F., Curt.* [*F.* Cerambycidæ. *G.* Clytus, *F.*]

Elongate, convex, black, thickly clothed with a greenish yellow pile, each elytron with four black spots. Head punctate; forehead with central furrow ; antennæ scarcely half the length of the body. Thorax oblong, narrowed and margined in front; sides faintly rounded. Elytra wider than the thorax, narrowed from base to apex, which is truncate and armed externally with a minute spine ; each with four denuded black spots, the 1st and 2d disposed transversely at the base; the other two on the disc. Underside and legs covered with greyish pile. Length, 5–5½ lines. " Found upon a window in Norwich." Probably imported in timber, and not truly indigenous.

Fig. 234. MESOSA NUBILA, *Oliv.* [*F.* Lamiidæ. *G.* Mesosa, *Serv.*] (Lamia nubila, *C.*)

Oblong, convex, black, variegated with ochreous and grey pubescence, and strewn with shining black punctures. Head as wide as the thorax in front, triangularly impressed between the antennæ with a fine deep line down the middle, and four black stripes on the vertex; antennæ longer than the body, slender, brown, each joint with a belt of grey pile at the base, and fringed beneath with fine long grey hairs. Thorax transverse, narrowed and finely margined in front, rounded at the sides ; base margined ; the disc with four black stripes. Elytra wider than the thorax; each with an oblique patch of grey pubescence on the outer margin a little before the middle, bordered before and behind with a more or less interrupted zigzag black band, the disc with three indistinct ridges. Underside covered with a dense grey pile. Legs variegated with yellow and black; tarsi black. Length, 4½–6 lines. On oaks and willows in the south.

Fig. 235. NECYDALIS MINOR, *L.* [*F.* Lepturidæ. *G.* Necydalis, *L.* (Molorchus minor, *Curtis.*)

Elongate, pitchy black, with a fine erect grey pubescence; mouth, antennæ, elytra, and legs reddish brown. Head as wide as the thorax, thickly punctate; antennæ setaceous, half as long again as the body, 12-jointed in the males, shorter and 11-jointed in the females. Thorax elongate; apex and base with the margin reflexed; sides obtusely tuberculated behind the middle; thickly punctate; with a depression in the middle behind, inclosing a small raised glabrous space, and having on either side a small oblong prominence. Scutellum triangular, thickly clothed with a silvery grey pubescence. Elytra wider and about one-fourth longer than the thorax, sparsely and rather coarsely punctate; each with a furrow within the humeral callus extending obliquely to the suture, and a small oblique, raised, yellowish white, smooth line in the middle, near the apex. Wings extending beyond the apex of the abdomen; pale brown, slightly iridescent. Length, 3–5 lines. Rare: in old hedges, and on flowers.

Fig. 236. SAPERDA TREMULÆ, *F.* [*F.* Lamiidæ. *G.* Saperda, *F.*] (S. Atkinsonii, *C.*)

Elongate, black, with a short, dense, greenish yellow pile. Head nearly as wide as thorax, convex, with a slender central line; antennæ setaceous, as long or a trifle longer than the body, black with greenish ashy pile, and fringed beneath with fine grey hairs. Thorax cylindrical, narrowly margined at the base, with a fine central channel. Elytra wider than the thorax; each with a row of four velvety black spots, one or more of which are occasionally wanting, on the disc. Legs covered with a greenish pile, intermixed with fine grey hairs. Length, 6–7 lines. The only example recorded as British was taken in a garden at Grove End.

Fig. 237. MONOHAMMUS SARTOR, *F. Curt.* [*F.* Lamiidæ. *G.* Monohammus.]

Elongate, brownish black, with a brassy tint, and a sparse yellow pubescence, shining. Head rugulose-punctate, with an impressed transverse line in front, and a central channel between the antennæ: antennæ slender, nearly twice the length of the body in the males, and about half as long again as the body in the females, black at the base, brownish towards the apex. Thorax transverse, sides armed with a large spine, transversely wrinkled in front and behind, the disc thickly and coarsely punctate, the anterior and posterior margins with a fringe of golden yellow hairs. Scutellum densely clothed with pale yellow pubescence. Elytra coarsely rugulose-punctate at the

PLATE XXVII.—*Continued.*

base, more finely so towards the apex; each with three indistinct ridges. Legs long, compressed, intermediate tibiæ, with a small tooth on the external margin towards the apex; basal joint of posterior tarsi distinctly longer than the second. Length, 12–16 lines.
Very rare. London; Norfolk; Devonshire; and near Manchester.

Fig. 238. ASEMUM STRIATUM, *L.* [*F.* Cerambycidæ. *G.* Asemum, *Eschscholtz.*]
(Callidium striatum, *Curtis.*)

Elongate, black, opaque ; clothed with a short yellowish grey pubescence. Head finely punctate, with a broad impression between the antennæ, a deep transverse one in front, and a shallow one on the crown; antennæ a trifle longer than the head and thorax, robust. Thorax sub-orbicular, slightly emarginate in front, sinuate truncate behind, the disc with a broad shallow central impression, and one or two ill-defined foveæ on each side of it; thickly granulate, with fine punctures between the granulations. Elytra densely and finely granulate, each with five or six ridges, the third and fifth usually most prominent. Legs short. Length, 5–9 lines.
Occurs in fir-woods in Scotland.

Fig. 239. RHAGIUM INQUISITOR, *Fab., Curt.* [*F.* Lepturidæ. *G.* Rhagium, *F.*]

Elongate, black, pilose. Head clothed with a dense yellow pile, with a denuded punctate space behind each eye, a transverse impression in front of the antennæ, and a fine central line on the crown; antennæ a little longer than the head and thorax, black at the base, brown towards the apex, covered with yellowish grey pile. Thorax elongate, constricted at the base and apex, with a broad transverse furrow in front and behind; sides rounded, with a large acute tooth in the middle; coarsely punctate, with an indistinct central furrow: clothed with a yellow pile. Elytra coarsely and irregularly punctate; mottled with a dense yellow pile; each with two transverse reddish yellow bands, the first just before, the second a little behind the middle, and between them, on the external margin, a large triangular denuded black patch : the disc with two ridges. Legs black, with a yellowish grey pile. Length, 6–11 lines. Common in dead firs, oaks, and other trees.

Fig. 240. OBRIUM CANTHARINUM, *L., Curt.* [*F.* Cerambycidæ. *G.* Obrium *Latr.*]

Elongate, ferruginous, tips of mandibles, base of antennæ and legs pitchy black, clothed with yellow hairs, shining. Head faintly punctate, with a large impression in front: antennæ slender, pubescent, basal joints pitchy black, ferruginous towards the apex. Thorax sub-cylindrical, constricted and transversely impressed near the base and apex : sides with an obtuse tubercle a little before the middle; sparsely punctate; with a short, broad, smooth, central elevation in front of the posterior impression. Scutellum obtusely triangular, traversed by a central furrow. Elytra as wide again and nearly thrice as long as the thorax; coarsely punctate, the punctures diminishing in depth towards the apex. Legs long, slender, clothed with long grey hairs; thighs swollen at the tips. Length, 3–5 lines. Rare. Near Brighton; Broxbourne, &c.

Fig. 241. STRANGALIA QUADRIFASCIATA, *L.* (Female.) [*F.* Lepturidæ. *G.* Strangalia, *Serville.*] (Leptura apicalis, *Curtis.*)

Elongate, black, with yellow and brown pubescence, elytra with four transverse yellow bands. Head elongate, constricted behind the eyes, thickly punctate, with a transverse impression in front, and a fine central impressed line: antennæ 11-jointed, black, pubescent, the three or four apical articulations reddish yellow. Thorax longer than broad, constricted and transversely impressed in front : base bi-sinuate, with a deep, transverse furrow : posterior angles prominent, acute; sides rounded in front : thickly punctate, with a faint central impressed line, and a large fovea on each side behind. Elytra nearly half as wide again, and more than thrice as long as the thorax: obliquely emarginate at the tips; thickly punctate: each with four transverse reddish yellow waved bands extending from the lateral margin to the sutural stria: the 1st near the base; the 2d oblique, at one-third from the base; the 3d at two-thirds from the base; the 4th nearly circular, just within the apex. Legs elongate, black. The males have the antennæ longer, and black throughout, and the thorax with the lateral foveæ obscure or wanting. Length, 7–9 lines. Widely distributed, but not common.

PLATE XXVIII.

Fig. 242. CALOMICRUS CIRCUMFUSUS, *Marsh.* (Male.) [*F.* Galerucidæ. *G.* Calomicrus, *Steph.*] (Luperus Brassicæ, *Curtis.*)

Oblong, above pale yellow, shining, base of thorax, scutellum, suture and external margin of elytra, and underside, black. Head minutely and sparsely punctate, with a transverse line between the eyes; antennæ black, the three basal joints yellowish brown. Thorax transverse, rounded and narrowly margined behind, sides nearly straight, all its angles rounded; finely and sparsely punctate, with two impressions on the disc a little behind the middle; the basal half occupied by a large, anteriorly dentate, black patch. Elytra faintly and sparsely punctate. Legs black, the anterior with the tibiæ and apex of thighs reddish yellow. In the female the antennæ are shorter, and the elytra distinctly dilated posteriorly. Length, 1½–2 lines. Widely distributed and locally abundant; on the common furze. (*Ulex Europæus.*)

Fig. 243. PHYLLOTRETA OCHRIPES, *Curtis.* (Male.) [*F.* Halticidæ. *G.* Phyllotreta Foudras.] (Altica ochripes, *Curtis.*)

Elongate-ovate, black, shining; anterior legs, tibiæ and tarsi of posterior legs, and a central more or less interrupted vitta on each elytron, pale yellow. Head coarsely punctate; antennæ with joints 1–3 pale yellow, 4 and 5 black, the latter compressed and dilated, the remainder (6–11) dusky. Thorax transverse, convex, narrowed in front, rounded at the sides, coarsely punctate. Elytra ovate, coarsely punctate, each with a broad, central, pale yellow stripe, more or less interrupted in the middle externally. In the female the antennæ are simple, the four or five basal joints pale yellow, the remainder dusky. Length, 1¼ lines. Not rare in marshy places in the south.

Fig. 244. PHYLLOBROTICA QUADRIMACULATA, *L.* [*F.* Galerucidæ. *G.* Phyllobrotica, *Redtb.*] (Auchenia quadrimaculata, *Curtis.*)

Oblong, ferruginous, shining; eyes, posterior portion of head, and two spots on each elytron, black. Head with a transverse, arcuate, impressed line between the eyes, and a deep central furrow. Thorax transverse, sub-quadrate, slightly narrowed anteriorly, smooth, with a short central furrow in front and behind, and a large shallow impression on each side of the disc. Elytra sparsely and finely punctate, each with a circular black spot at the base, and a transverse black patch near the apex. Underside, save the prothorax, black, sparsely clothed with yellow hairs, thickly and finely punctate. Length, 3–4 lines. Common in marshy places in the south.

Fig. 245. LEMA PUNCTICOLLIS, *Curtis.* [*F.* Crioceridæ. *G.* Lema, *Fab.*] (Crioceris puncticollis, *Curtis.*)

Oblong, blue with a violet tint, shining; antennæ and tarsi black. Head coarsely rugulose-punctate, forehead with two minute glabrous tubercles, crown with an elongate fovea. Thorax a trifle longer than broad, constricted at the sides a little behind the middle, with a shallow transverse impression at the base, and two coarsely and roughly punctate depressions on the disc, the intervening space nearly smooth; coarsely and sparsely punctate at the sides. Elytra oblong; each with ten rows of deep punctures, interstices flat, smooth. Legs rather long, sparingly clothed with grey hairs. Length, 2–2¾ lines. Widely distributed, but rare.

Fig. 246. HÆMONIA EQUISETI, *Fab.* [*F.* Donaciidæ. *G.* Hæmonia, *Latr.*] (Macroplea Equiseti, *Curtis.*)

Oblong, reddish yellow; head, antennæ, and underside, black, with a fine short silky grey pile. Head with a broad, deep furrow in front between the eyes; mouth ferruginous. Thorax elongate, rounded in front, the anterior angles obtuse but slightly prominent; posterior angles acute and prominent; sides with an oblong tubercle before the middle, thence gently

55

PLATE XXVIII.—Continued.

sinuate to the posterior angles; very finely shagreened; with a fine interrupted and abbreviated central impressed line, and an oblique linear black mark on each side. Elytra truncate at the tip, the external angles produced into a long acute spine, the internal or sutural angles obtuse; each with ten deeply punctate striæ, and one, abbreviated, at the base next the suture, the punctures black; interstices smooth, convex, the 2d, 4th, and 8th, wider and more convex than the rest. Legs long and slender, ferruginous, apex of femora and tarsi, except the base of the fourth joint, black. Length, 3¼-4½ lines. Very rare.

Fig. 247. DONACIA TYPHÆ, *Brahm, Curtis.* [*F.* Donaciidæ. *G.* Donacia, *F.*]

Elongate, above brassy or brassy green, with a silky lustre, elytra with a coppery or steel-blue stripe next the suture. Head finely granulate, clothed with a dense silvery-grey pile, forehead with a broad central channel and a narrow one on each side; antennæ with the basal joint brassy, the remainder pitchy, their bases narrowly ferruginous. Thorax a little longer than broad, slightly narrowed behind; anterior angles obtuse, but slightly prominent, deflexed; sides with an oblong tubercle immediately behind the anterior angles, succeeded by a minute emargination, beyond which they are sub-angularly produced, and thence sinuate to the base; posterior angles obtuse, slightly prominent; the disc rather depressed; thickly punctate; with a shallow central furrow. Scutellum clothed with a silvery grey pubescence, with a narrow central denuded space. Elytra emarginate at the tips; striate-punctate, the punctures irregular at the base and apex; interstices slightly convex, very finely transversely rugulose. Legs long, slender, brassy black, pubescent, base and inner edge of tibiæ and base of thighs ferruginous. Length, 4-1¾ lines. Not common, although widely distributed.

Fig. 248. PSYLLIODES CHALCOMERA, *Illig.* [*F.* Halticidæ. *G.* Psylliodes, *Latr.*]
(Macrocnema unimaculata, *Curtis.*)

Ovate, convex, above dark blue, sometimes with a green or violet tint, shining. Head minutely punctate, the mouth, and antennæ, pitchy, four basal joints of the latter ferruginous. Thorax transverse, convex, narrowed in front, rounded at the sides, thickly and minutely punctate, with large irregularly scattered punctures. Elytra wider than the thorax, ovate, punctate-striate, insterstices flat, minutely punctate. Beneath black with a faint brassy tint. Legs ferruginous, posterior thighs much swollen, brassy black, the extreme tips ferruginous. Length, 1-1¾ lines. Not uncommon; on thistles.

Fig. 249. GALERUCA VIBURNI, *Payk., Curtis.* [*F.* Galerucidæ. *G.* Galeruca, *Geoffr.*]

Oblong, convex, reddish brown, finely shagreened, minutely punctate, with a dense silky yellowish grey pile, sub-opaque. Head with a large triangular black patch on the crown; antennæ ferruginous, the tips of each joint fuscous. Thorax transverse, narrowed in front, base tri-sinuate, sides rounded; with a broad, shallow, central channel, and a large impression on each side, the channel and the lateral margins black. Elytra dilated posteriorly, convex, the shoulders, and not unfrequently the entire outer margin, black or fuscous. Length, 2¾-3 lines. Not uncommon.

Fig. 250. CASSIDA OBLONGA, *Illig.* [*F.* Cassididæ. *G.* Cassida, *L.*] (C. salicorniæ, *C.*)

Oblong ovate, convex, slightly shining, above pale green, fading to greyish yellow, with a faint green tinge after death. Head and underside black; abdomen with a broad testaceous margin; antennæ and legs testaceous, the former fuscous towards the apex. Thorax transverse, narrowed anteriorly, base bi-sinuate, rounded at the sides and in front; posterior angles well developed, slightly acute, summits obtuse; sides flattened, somewhat deflexed, thickly and minutely reticulate, and rather thickly punctate; disc nearly smooth. Elytra punctate-striate; interstices flat, smooth, the 2d and 3d wider than the rest, the latter with a short irregular row of large shallow punctures in the middle; each, during life, with a broad, brilliant, silvery green stripe occupying the space between the 2d and 5th, occasionally extending to the 7th stria. Length, 2¼-2¾ lines. Not uncommon in marshy places.

56

PLATE XXIX.

Fig. 251. LYCOPERDINA BOVISTÆ, *Fab., Curtis.* [*F.* Endomychidæ. *G.* Lycoperdina, *Latr.*]

Elongate ovate, pitchy black or chestnut, shining. Head finely and rather thickly punctate,
smooth in the centre, with a longitudinal impression; antennæ with the 2d and 3d joints little
more than half the length of the 1st, about twice as long as broad, 4–8 sub-globose, 9–11
oblong, forming an indistinct club. Thorax sub-quadrate, narrowed behind, sides rounded in
front, straight posteriorly; very finely and sparsely punctate, with a deep impressed longitudinal
line on each side behind. Elytra ovate, very finely and sparsely punctate, each with an entire
sutural stria. Legs pitchy red. Length, 2–2¼ lines.

Not common. Inhabits the common Puff-ball (*Lycoperdon Bovista*). Birchwood, Kent;
Guildford, Sanderstead, and Wimbledon Park, Surrey; &c.

Fig. 252. COCCIDULA SCUTELLATA, *Herbst.* [*F.* Scymnidæ. *G.* Coccidula,
Kugelaun.] (Cacicula scutellata, *Curtis.*)

Oblong, ferruginous red, slightly shining, clothed with a short silvery grey pubescence, eyes
and scutellum black, elytra with a large triangular black patch at the base, an ovate spot on the
external margin a little behind the middle, and a circular one at about one-third from the apex
near the suture, sometimes united to the lateral spot and forming a flexuous transverse band.
Head thickly and rather finely punctate; antennæ and palpi reddish yellow, the club of the
former dusky. Thorax transverse, rounded and narrowly margined at the sides and in front
behind the eyes, thickly and rather finely punctate. Elytra as thickly but not quite so finely
punctate as the thorax, and with irregular longitudinal rows of larger punctures. Legs
ferruginous. Length, 1½–1¾ lines.

Not common. Inhabits marshy places amongst reeds. Banks of the Thames below Gravesend; Southend; Plaistow Marshes; Holme Fen, Hunts; Horning Fen, Norfolk; near Hammersmith; &c.

Fig. 253. CRYPTOCEPHALUS BIPUNCTATUS, *Linn. var.* [*F.* Cryptocephalidæ.
G. Cryptocephalus, *Geoffr.*] (C. bipustulatus, *Fab., Curtis.*)

Oblong sub-cylindrical, black, shining, parts of mouth and basal joints of antennæ pitchy,
elytra with a large reddish yellow patch within the apex. Head nearly smooth behind, somewhat
coarsely and rather thickly punctate in front, with a scant short yellowish pubescence. Thorax
transverse, very convex, sub-globose, narrowed in front, sides narrowly margined and slightly
rounded anteriorly, more widely margined and nearly straight behind, exceedingly minutely and
sparsely punctate. Scutellum oblique, elongate, triangular, rounded at the apex, smooth, with a
shallow oblong impression at the base. Elytra sub-cylindrical, with a conspicuous longitudinal
impression within the shoulders, and a small raised space round the scutellum; coarsely punctate-
striate, interstices flat with a few transverse wrinkles. Length, 3–3½ lines.

Rare. Lyndhurst, Hants; Lancashire.

Fig. 254. PRASOCURIS BECCABUNGÆ, *Illig.* [*F.* Chrysomelidæ. *G.* Prasocuris,
Latr.] (Helodes Beccabungæ, *Curtis.*)

Elongate, rather depressed, deep blue, shining, apex of abdomen reddish yellow. Thorax sub-
quadrate, sides rounded in front, coarsely and sparsely punctate. Elytra more than thrice the
length of the thorax, sides parallel; rather finely punctate-striate; interstices flat, minutely
rugulose punctate. Length, 1¾–2¼ lines.

Common in ditches and brooks on the short-leaved Brooklime (*Veronica Beccabunga*), and
other plants.

Fig. 255. LABIDOSTOMIS TRIDENTATA, *Linn.* (Male.) [*F.* Clythridæ. *G.* Labidostomis, *Lacord.*] (Clythra tridentata, *Curtis.*)

Elongate, rather convex, bluish green, shining, parts of the mouth pitchy, base of antennæ
and elytra testaceous yellow. Head rugulose in front, thickly punctate behind, forehead with a
broad shallow depression. Thorax transverse, anterior margin broadly emarginate, base bi-sinuate;
sides nearly straight in front, rounded behind, posterior angles prominent; rather minutely and
very thickly punctate. Elytra nearly thrice as long as the thorax, rather thickly and somewhat

57

PLATE XXIX.—*Continued.*

coarsely punctate, each with two or three indistinct longitudinal ridges. The female differs in being rather larger, and in having the anterior legs shorter. Length, 3¾-4½ lines.
Found on hazel, sallow, and birch, at Darenth Wood, Kent; Coombe Wood, Surrey; in Worcestershire; Yorkshire, &c.

Fig. 256. MANTURA MATHEWSII, *Curtis.* [*F.* Halticidæ. *G.* Mantura, *Steph.*]
(Cardiapus Mathewsii, *Curtis.*)

Oblong-ovate, convex, above deep blue, or green with a brassy tint, shining, legs yellowish red. Head rather coarsely and sparsely punctate; antennæ black, the apex of the 1st, and the 2d, 3d, and 4th joints reddish yellow. Thorax transverse, narrowed behind, base bi-sinuate, sides rounded; rather coarsely and thickly punctate, with a short oblique impressed line on each side at the base. Elytra sub-cylindrical, deeply and rather coarsely punctate-striate; interstices flat, finely rugulose. Beneath black, finely and rather sparsely punctate. Legs yellowish red, thighs pitchy black. Length, 1-1¼ line.
Not common, although widely distributed in the south and midland counties.

Fig. 257. ENTOMOSCELIS ADONIDIS, *Fab.* [*F.* Chrysomelidæ. *G.* Entomoscelis, *Redtb.*] (Chrysomela Adonidis, *Curtis.*)

Oblong-ovate, convex, above bright yellowish red, anterior portion of head, and a spot on the vertex, antennæ, a large discoidal patch on the thorax, scutellum, a longitudinal stripe next the suture, and another on the disc of the elytra (sometimes extending nearly to the lateral margins), black. Thorax transverse, narrowed in front, sides faintly rounded, posterior angles rectangular; finely and rather thickly punctate. Elytra finely, rather sparsely and irregularly punctate. Underside and legs black. Length, 3¼-3½ lines.
Supposed to have been taken, many years since, in Lincolnshire.

Fig. 258. ENDOMYCHUS COCCINEUS, *Linn., Curtis.* [*F.* Endomychidæ. *G.* Endomychus, *Panzer.*]

Ovate, moderately convex, bright scarlet red, shining, head, a patch on the disc of the thorax, scutellum, two spots on each elytron, and the underside of the thorax black. Head finely and sparsely punctate, with a transverse impression between the eyes: parts of the mouth ferruginous; antennæ black, the two basal joints pitchy, tip of apical joint ferruginous. Thorax transverse, narrowed in front, sides nearly straight, base bi-sinuate, posterior angles acute; finely and sparsely punctate, with a fine transverse impression near the base, and a deep impressed longitudinal line on each side behind. Elytra ovate, finely punctate. Legs black or pitchy. Length, 2-2½ lines.
Local, but widely distributed throughout the country. It inhabits beneath bark and in decaying stumps of trees, especially such as nourish fungi, and frequently occurs in profusion.

Fig. 259. HALYZIA OCELLATA, *Linn.* [*F.* Coccinellidæ. *G.* Halyzia, *Mulsant.*]
(Coccinella ocellata, *Curtis.*)

Short-ovate, convex, shining. Head black, rather finely punctate, parts of the mouth and antennæ pale yellow, the latter fuscous at the apex. Thorax transverse, emarginate in front, rounded at the base and at the sides, finely punctate, black, the lateral margins broadly, and two spots at the base before the scutellum, yellowish white. Scutellum black. Elytra yellowish red, finely and thickly punctate; each with nine black spots encircled with pale yellow, the 1st, ovate, oblique, on the shoulder; the 2d, linear, close to the scutellum; the 3d, ovate, on the outer margin at about one-third from the base; the 4th, ovate, on the disc at about one-fourth from the base; the 5th, ovate, near the suture at about one-third from the base; the 6th, 7th, and 8th, ovate, placed transversely and rather obliquely at about one-third from the apex; the 9th, linear, situate obliquely near the apex, midway between the lateral margin and the suture. Underside black, posterior margins of abdominal segments reddish yellow. Length, 4-5 lines.
Widely distributed, but not common, in fir woods.

London:—STRANGEWAYS & WALDEN, Printers,
28 Castle Street, Leicester Square.

INDEX.

a...ta...chrome   ...eetus fulvicorn...   Ce...cus ...

...thrus rostrat...s   Calosoma sycophanta   ...d...s c...g...

...ndia la...ret...   ...m s...um se...r...   ...

... .

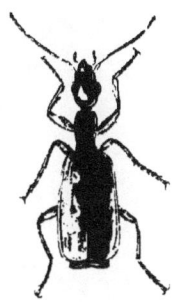

I ... fasciatus

_____ _____   Cychrus elongatus   Nebria Burrellii.

Dyschirius inermis   Leistus Reachii   Pterops_____

... depressus ... ...

Callistus ... ... Chlaenius ... ...

... ...

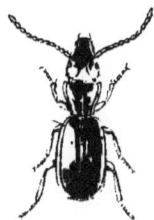

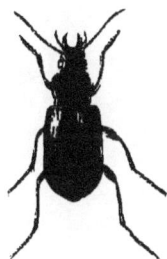

 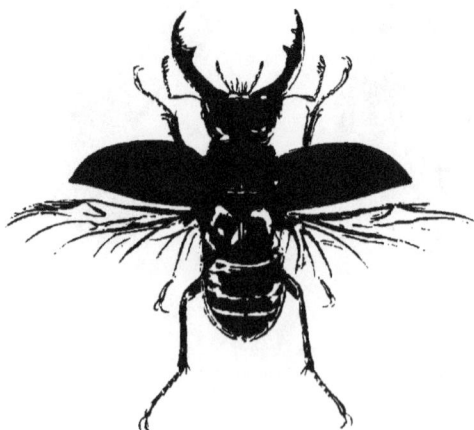

81

82

83

Platycoria coralades.

Ips depressictus.

Micropop intersecula

84

85

86

Hydroia chalcenema

Lernoaa onneops

Chara solen.

89

Phiotagr

99.

100.

101.

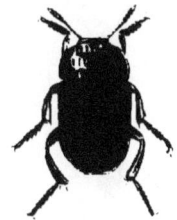

Catops assimilatus.

Necrodes lateralis.

Silpha opaca.

102.

103.

104.

. . . . .

. . . .

. . . .

. . .

1 . .

107.

. . . . .

. . . .

. . . . . .

Tachnus rufipes

Quedius laterale

Philonthus marginatus

Tachus fucicola

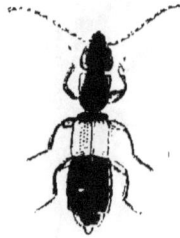

Achenium depressum

Lathrobium terminatum

Dactylosternum hemisphaericum

Phytosus spinifer   Phytosus spinifer   Acylophorus glabricollis

120   121

Lathrobium terminatum   Lathrobium Spadix   Oxyporus rufus

123   124

Xantholinus punctulatus   Xantholinus collaris   Olisthaerus

Paederus fuscipes   Stenus Morio   Bogilus fragilis

128   129   130

Palagria thoracica   Pronius sulcatilis   Amaurobius puncticollis

131   132   133

II   ...   Lathanus elongatus   ...

  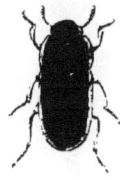

Tenebrio obscurus.

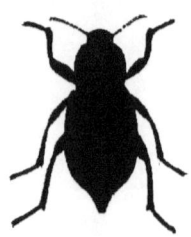

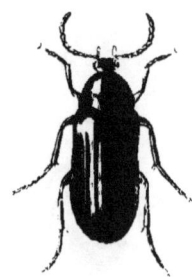

153

Mordella abdominalis

154

Sitaris humeralis

155

Meloe brevicollis

156

Nothus bipunctatus

157

Conopalpus testaceus

158

Dictiona flavipes

159

Cantharis vesicatoria

160

Edemera sanguinicollis

161

Pyrochroa coccinea

162

163

164

Hylecoetus dermestoides

Lymexylon navale

Anthicus thialis

165

166

167

Abbrum scutas

Ptinus fluripunctus

Mornum caustum

168

169

170

Anobium paninax

Rhipidius subsus

Serrocerus pectinatu

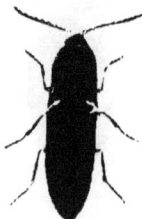

Corixa sahlbergi

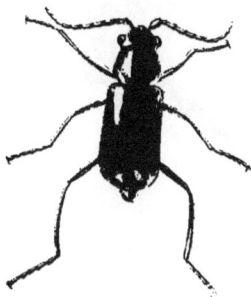

Necrodes surinamensis

Staphylinus erythropterus

Plate 14

. . . . Nemosc . . . . gabora . . . . mus

101

. . . . barus . . . . . .

103 2 4 106

. . . .

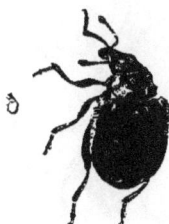

Magdalis armicollis.　　　　Rhynchites.　　　　Attelabus rhois.

Balaninus nasicus.　　　　Balaninus uniformis.　　　Hylobius pales.

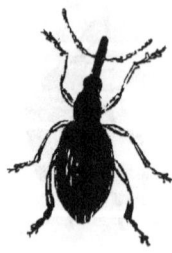

Pissodes strobi.　　　　Arrhenodes.　　　　Polydrusus impressus.

Bruchus

anisandra muricata

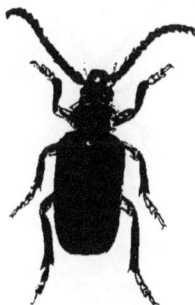

Hetero[...]icornis.   [...]mia [...]rnia.   Molorchus [...]mer

256

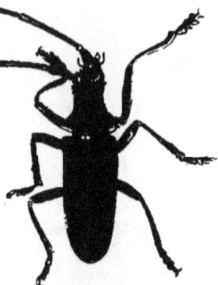

Saperda Atkinsoni.   M[...] Restonus saron[...]   C[...]

247

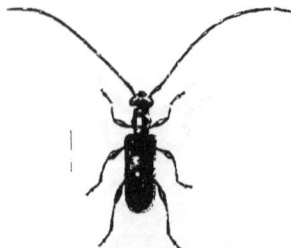

Fragmia in[...]uisitor   Obr[...]m [...]   [...]

[...]cidæ.

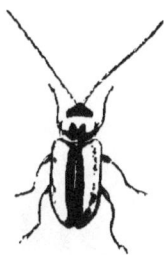

Colaspis brassicae   Disonycha xanthomelas   Diabrotica 4-maculata

245   246   247

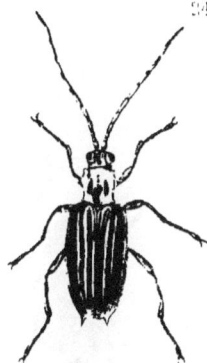

Chlaenius pruinicollis   Macroplea equiseti   Donacia typhae

248   249   250

27 68

Liocyanidra Komatae ... ... Hygroscybhus digums vn

165

Held for Bearchungor ... Hygdro ruddentara ... ardragma Mardier

115 266 299

monnoply nebotable ... Inver gdose menenon ... coreudelle vo le

* 9 7 8 3 3 3 7 1 4 4 0 4 3 *